My Life
with the Printed Circuit

My Life with the Printed Circuit

Paul Eisler

Edited with notes by Mari Williams

Bethlehem: Lehigh University Press
London and Toronto: Associated University Presses

© 1989 by Associated University Presses, Inc.

Associated University Presses
440 Forsgate Drive
Cranbury, NJ 08512

Associated University Presses
25 Sicilian Avenue
London WC1A 2QH, England

Associated University Presses
P.O. Box 488, Port Credit
Mississauga, Ontario
Canada L5G 4M2

The paper used in this publication meets the requirements of the American National Standard for Permanence of Paper for Printed Library materials Z39.48-1984.

Library of Congress Cataloging-in-Publication Data

Eisler, Paul, 1907–
    My life with the printed circuit.

    Bibliography: p.
    Includes index.
    1. Eisler, Paul, 1907–   .   2. Electric engineers—
Great Britain—Biography.   I. Williams, Mari.
II. Title.
TK140.E37A3   1989   621.3'092'4 [B]   88-81397
ISBN 0-934223-04-1 (alk. paper)

*To the memory
of my dear late wife Frieda*

Contents

Preface

As I approached the end of a long and fruitful professional career I felt increasingly urged to write an account of my experiences as an inventor. I have written this book as an autobiography of my experiences as a freelance inventor. It is a behind-the-scenes report of my attempts to get my ideas through the jungle of bureaucratic obstacles. It is the tale of my life-long fight as an individual facing the obstructions of established powers and groups. It offers, moreover, a look at the patent system from the point of view of an inventor, giving a distinctly different picture from that provided by the big organisations, or by patent attorneys. Inventions are the base of development and my experience has convinced me that the patent system is failing to help the inventor. It should be a priority of Parliament to improve the system, and I trust that my autobiographical account will give factual fuel for that purpose.

The book is the story of my inventions since I came to England in 1936. Among my many inventions the best known is the Printed Circuit, which has become the basis of the modern electronics industry and thereby increasingly affects everyone's day to day life. It was introduced some 40 years ago in proximity fused anti-aircraft shells which played such a decisive role in the defeat of Germany in the Second World War. An appreciation of the significance of my Printed Circuit was recently expressed by a major supplier of laminates for circuit boards, in a calendar whose theme was the great inventors of electrical technology. In this calendar I appear in a galaxy together with Volta, Faraday, Bell, Edison, Marconi and others.

My book describes the bureaucratic impediments which accompanied the attempts to put my ideas into practice. It is a tale of a lonely struggle to guide my inventions past the internal rivalries of large corporations; it is particularly these experiences that I wish to be made known, especially to anyone interested in promoting his own inventions.

My desire to write this book was strengthened by my failure to find a similar account. I felt it my duty to highlight some of the flaws in the organisations on which our prosperity so crucially depends. My

9

Photograph of the author with the first radio set using a printed circuit chasis and aerial coil. *(Photo courtesy of Maurice Hubert, Multitech [UK])*

observations and experiences of these matters should thus be of considerable interest to a wide audience, particularly to those interested in the workings of our society.

I particularly wish to thank Mrs. Sarah Manson for her unfailing support and encouragement in bringing this book to publication.

My Life
with the Printed Circuit

1: The Beginnings in Vienna

When, at the age of 23, I received my Diploma in Engineering from the Technical University at Vienna, I was not especially interested in inventing or in inventions. All I wanted was a job as an engineer active in the production of electrical goods.

Alas, the facts that I had good references, was keen, and had only modest salary requirements were of no avail. The year was 1930 and the country, Austria.

The electrical industry was already threatened by the German Nationalists.[1] Indeed all jobs for young engineers were reserved for members of the virulently antisemitic German-Nationalist Student corporations whose Jew-baiting excesses within the walls of the University had been a regular feature of my student life. For I was a Jew; I was not at all religious but I was proud to be Jewish and recognisable as such without the slightest doubt. Even my name indicated it.

So, whilst my applications for jobs were promptly returned without even the chance of an interview, all the non-Jewish students who had graduated with me obtained suitable jobs. In view of the political developments in Germany and Austria I had expected this and was not at all surprised, depressed, or daunted. Instead I acquainted myself with the young, fast developing radio trade.

By accident, I heard that the English firm His Masters Voice (HMV), famous for their records, was seeking an engineer to go to Belgrade for a special job. At that time HMV, still had a centre in Vienna for their trade with the countries of the old Austro-Hungarian Empire and the Balkan States. Their agent for the kingdom of Serbia had a shop in the centre of Belgrade selling their records, radiograms and other lines imported directly from Hayes in England; he also acted for them in any allied business. If I accepted the job with HMV, I was told that he would receive me, give me the details of what was expected, and generally look after me. At the age of 23 and in a politically precarious position I looked at this offer as a lucky break: it would be an adventure at the same time as being gainful employment. I accepted at once and very soon boarded the train to Belgrade. There I experienced my first insight into the workings of business as one of its junior officers.

15

My job was to develop equipment for eliminating the heavy disturbance of radio reception and sound transmission in railway trains. Broadcasting music from a radiogram in one carriage to all other parts of the train would be very lucrative if every passenger could be charged for the hire of headphones—early precursors of today's Sony Walkman. Many travellers were illiterate and even those who were able to read found it too much of a strain in the consistently poor light provided. There was therefore an obvious market for other forms of diversion from the tedium of travel. But any sound system had to be as free as possible from distortion or interference. HMV was competing with the German firm Koerting and there was considerable business advantage to be gained by the firm finding the most economic solution. I was employed by HMV to do this. It was known that a reasonable system existed on the provincial line from Belgrade to Sarajevo in Bosnia, but the trains running along it were primitive with no electric lighting and hence no dynamos to cause interference. I was to face more challenging problems. In the first instance I worked on trains running from Belgrade to Niś, on which the elimination of disturbances from the dynamos on the more modern main line carriages might permit the introduction of an even larger business.

I understood what was required of me; and I also understood that I should be responsible for the repair wherever possible of the defective radiograms and radios accumulating in the shop in Belgrade. The agent there was to get me the necessary labour permit and to provide me with facilities for the inspection of gear and of wagons. It was clear that I would also need to carry out a variety of tests on new equipment. All that was set up within a few weeks by which time I was crawling underneath railway carriages and making friends with engine drivers who understood German. But working at a distance from the head office of a firm is always difficult and the facilities with which I had been provided were far from satisfactory. Specifications often had to be made more by guesswork than by precise measurement as the instruments I could get hold of were quite inadequate for the job.[2] I ordered chokes and capacitors from Hayes but had a long wait before they arrived. However, when they did I was able to put them together and with the help of a railway technician fixed them onto one carriage. Eventually, despite delays and problems of communication, the great finale came. The carriage—with half a dozen railway officials, our agent, and myself as passengers—was coupled to an engine and we drove off. First we travelled along sidelines and then—the final demonstration—full speed towards Niś. I had guessed, or estimated, correctly and the interference was, even at its worst, reduced to an acceptable level. We returned triumphantly and the agent reported in style.

Subsequent developments were, however, quite unexpected. HMV quote for the equipment (radiograms, microphones, headphones, filter chains, and library of records) was accepted by the Serbs but payment was offered only in grain, not pounds sterling. The foreign exchange crisis in Serbia had reached new heights;[3] but Hayes would not play ball. They refused to become traders in grain and broke off negotiations. That was the end of my prospects with HMV and in due course I returned to Vienna. I was deeply affected by the social, political, and human conditions which I had witnessed from close range day and night, and which killed the last remnants of the socialism left from my teens.

In Vienna there was, of course, no job available for me as an engineer. As I had to earn a living somehow, I started writing for papers and founded—together with other young unemployed technicians—a weekly radio journal for which I designed a coordinate arrangement for the radio programmes. During this time I had to study basic printing technology and I assisted actively in the printing shop in order to integrate my layout with the standard inexpensive printing techniques to which the shopfloor printers were used. The coordinate design proved a success with the public and our radio journal was taken over after a few months by the big social-democratic publishing house Vorwärts. It was in this way that I got an editorial post and the printing experience which was to prove of such relevance later; moreover, at the time what was far more important for me, I received my first monthly salary.

This relative prosperity lasted only until February 1934, when the putch of the Austrian fascists spelled the end of all social-democratic enterprises. It was also the end of journalism for me, and the beginning of another period of great uncertainty as the hunt for people like former Vorwärts journalists—a classification applying to me as well—got under way. It was some months before that subsided and the worry of how I might earn a living could once again come to the foreground.

Under the new Austro-fascist regime it would have been almost impossible, and certainly dangerous, to reorganise a collective of a few friends to start any new enterprise, and particularly anything to do with publishing. Any such group would at once have been suspected of belonging to the underground, whether or not it actually had any political motivation. All I could do was to work alone in circumstances designed not to attract the attention of the new authorities. Indeed what I really had to do was to consider emigration to America or Britain.

With no relatives or friends living in either place, nor indeed in any other country abroad my problem was to get a visa. It was my attempts to solve this which made me take a serious interest in patents for the first time. Thus it was that my early involvement in invention and particularly in the system meant to protect and stimulate inventors, resulted not from any technical or financial motivation but from political expediency.

It occured to me that a letter of invitation from a well-known, highly reputable firm might help me to gain entry into France or Britain. One way to obtain such an invitation would be to ask the firm whether they would be interested in an invention of mine in a field in which they were prominent. Moreover, although I was not especially pursuing new inventions in an organised way there were two, rather different, inventions for which I had made patent applications in Austria; I therefore decided to try to interest foreign companies in them.

The first had emerged from my work for a doctorate at the Technical University. It was on graphical sound recording. When I wrote about it to Thomson-Houston in Paris, they claimed not to be interested. However, they recommended me to their British sister firm who sent me a letter inviting me to visit them for discussions. I thus had the means of convincing immigration officials in Britain that my arrival in their country was in response to an invitation.

My position was further enforced through the second patent application. This was for an idea which, at the time, seemed of no possible practical value, namely Stereoscopic Television. It was an idea that I had by accident, after a friend showed me glass plates with screens of vertical lines behind which there were photographs visible slightly differently to the right and left eye. The concept was used as an advertising novelty in shop windows; but I was not interested in advertising. I saw at once the possibility of developing a stereoscopic television system using vertical instead of the horizontal lines in present-day television. Around that time engineering related to television of any sort was still a very new area of expertise with many technical problems, one of which was the width of the band of modulation frequencies.[4] The system of vertical line scanning which I proposed would use a much narrower frequency band, and this principle, because it was so basic, proved to be of interest to the Marconi-Wireless Telegraph Company.[5] In the event prosecution of the patent later revealed that I was the second man to have had the general idea, the first being the man who became chief executive of the Radio Corporation of America; mine was, however, a substantial improvement over his and led to a number of other important theoretical developments. It had moreover aroused the interest of another British-based firm and so greatly helped me in my quest to leave Vienna.

In 1936 then, armed with two patent applications, I left the city of my birth and made my way to an unknown land where, as it turned out, I was to spend the rest of my life.

2: Early Experiences in England

Of the two patent applications I had brought with me to England it was, surprisingly, the second and obviously less practicable that was of greater immediate use to me. The Marconi-Wireless Telegraph Company was encouraging in its response to my ideas for a system of stereoscopic television. On my arrival in England the company, was ready to reach an agreement with me. Whatever the executives' motives, they appreciated that my patent promised the greatly sought reduction of frequency modulation band width. They bought the patent rights outright, with the minimum of fuss. They payed me the then generous sum of £250, certainly sufficient to provide for my needs for several months.

But although this money was soon forthcoming, in supplying me with a measure of financial support, I was not allowed to take a job in Britain. I was thus faced with the problem of securing some sort of income, and turned to invention as a possible means of so doing. At the time I was living in a very small room in a London boarding house in Hampstead and was consequently very restricted in what I could attempt experimentally. I was trained in engineering with work experience in electrical circuitry; the necessary tools and implements for experimentation in this branch of engineering were sufficiently small and inexpensive to allow me to buy enough and to use them in my small room. Thus it was that I turned my attention to inventions within the realm of electrical circuit boards. To this I was able to bring the other practical expertise I had developed: that of printing technology acquired during the years spent in Vienna as a technical editor. It was this combination which led me to work on the most reputed and significant of my inventions: the printed circuit.

Long before the First World War a telecommunications industry existed in the developed states of Europe and in the United States, and quite complex electrical circuits were mass produced in that industry.[1] In addition there was a radio industry which had also been mass producing complicated electrical circuits since the 1920s. Both these industries, I believed, would be interested in adopting a method of circuit production based on the semi-automatic process of printing, instead of having them wired by hand labour. The two industries comprised many firms, large and small, all claiming progressive attitudes, all in competition with cash offer and all, therefore, presumably interested in labour-saving technology. Moreover, all the materials, skills and machines for producing printed circuits were commercially available, ready and inexpensive.

My aim, therefore, was to produce a circuit board onto which strips of metal could be adhered using a printing process, and to find a readily understandable means of demonstrating this new technology. I decided to make a radio set using printed circuits instead of conventional wiring, and with primitive tools, instruments, and implements bought at a shop catering for radio amateurs, I put together a small two-valve set. Using a bakelite sheet as chassis, I hoped to replace all wires by metallic strips of maximum width on both sides of the sheet. After a few failures I eventually obtained satisfactory results and studied ways and means of printing the strips. Although I already had experience of printing from my days as technical editor of the radio weekly *Rundfunk*, I learned far more after reaching Britain, by installing myself in the library of the British Museum.

I became fascinated by the impressive technological achievements of the printing art. I saw this art as a whole: letterpress and gravure, lithography, offset and screen printing, engraving and photomechanical printing. As I read, I imbibed all the main processes like the wisdom of ultimate redemption.

There was no doubt in my mind that everything which could be drawn in black and white could be magnified to poster size or more, or else reduced in size to dimensions smaller than a postage stamp. It could be printed by any one of a dozen processes on copper or on other materials offering a very small or a larger resistance to the electric current. The flat, basically two-dimensional nature of these conductors could then offer new and so far undreamt of facilities for the whole electrical and electronics industry.

Whilst this industry had to produce ever more complex networks of electric circuits, mass production of these might become ever more fault ridden. Printing, however, was a recognised method of automatically reproducing great numbers of exact copies of a two-dimensional original, irrespective of its complexity. Printed Circuits had therefore clearly something to offer to mass producing firms. In 1936 the most prominent mass production in the electronic industry was that of radio receivers. For the time being this limited the possible practical application of the idea to the manufacture of these sets. Still, it constituted a promising start.

Invigorated by my study I spruced up the little radio set I had made by hand—painting, leafing, and ironing, tested it and tried to find an agent to get me an interview with a big radio manufacturer. I found one in a relatively short time who introduced me to Plessey.

I was taken to meet the director in charge of radio production at the Plessey company in Ilford, Essex. I demonstrated to him my first printed circuit invention in the form of a complete radio set which worked perfectly. On the visual and audible evidence produced in his office he could not very well refuse the invention for one or more of the reasons conveniently employed in the usual "throughly considered" judgments of the Research and Development departments of most large firms. He could not, for instance, praise the invention's ingenuity but doubt whether it would work at all or think that it would only do so after a prolonged and incongruously expensive period of development.

Neither could he doubt the novelty of the idea, as he had already stated, when accepting my offer to demonstrate my invention in confidence, that neither he nor his staff had ever heard that a radio set could be produced by printing its connections. I could imagine a number of other reasons often used for not taking up an outsider's proposal: the excuse being "Not Invented Here". On this score, acceptance might discredit the firm's own Research and Development department, or perhaps simply detract from the erstwhile dominant group in the firm, or even provide ammunition for another group in the power struggle within the firm.

I recall Chester Carlson whose invention of Xerography was rejected by twenty large organisations. Would I have to go through a similar stage? The Director of Plessey neglected the danger that I would or could successfully offer such a basic invention to one of Plessey's competitors. He obviously understood well the structure and spirit of the other big firms in the radio industry at that time. It seems that only politicians, lawyers, economists, patent agents and others outside the industry believe that there is eager competition by firms to take up inventions from independent outsiders.[3]

Nevertheless, to my chagrin, my invention was turned down. The reason given for refusal was unexpected, but it was one for which the firm could very easily claim authority. It was pointed out to me that the work which my invention would replace was carried out by girls and that "girls are cheaper and more flexible". I was, of course, deeply disappointed, but I accepted the verdict. I could hardly do otherwise.

My confidence in the idea was not weakened but I sensed that it may take time to find another interested party. With no work permit nor money enough to live for more than a short time, I could not afford the luxury of waiting for the idea to be taken up.

I returned to one of my earlier inventions—the patent on graphical sound recording, about which I had corresponded with Thomson-Houston.[4] I now made new representations to Thomson-Houston and they invited me to visit them in Rugby. I was received in a very friendly manner but camouflaged in patronising compliments, their response was in essence quite clear: they were more interested in other lines of recordings on sound film and they would not take up my method. However, during the ensuing discussion I learned about Odeon Theatres, a fast-growing new chain of cinemas which Thomson-Houston supplied with projection equipment. The people at Rugby offered to give me a recommendation, and this together with a personal introduction gave me the chance of interview with Oscar Deutsch, the founder and head of Odeon Theatres. By 1936 this comparatively new organisation was building 50 "Super Cinemas" a year all over England, each of which set new standards for cinemas. Its engineering division handled several hundred items ranging from seats to sound and projection equipment, and managed everything from large-screen television to the running a maintenance organisation.

My interview was a success. Deutsch engaged me for a trial period and obtained a work permit for me. Although the engineering division at Odeon Theatres was comparatively large it had no research and development department. I was to be a single man substitute for it and it would be up to me to prove the value of such an activity. My functions were those of an "ideas" man, an organiser of new technical features and a designer, inventor, and development engineer for new cinema equipment. I enjoyed the challenge. As a result I had no time to spare for printed circuits: other technical problems needed all my attention. Nevertheless my time in cinema was to prove very fruitful for my career as an inventor.

I was with Odeon Theatres during the building of the Odeon, Leicester Square, and assisted in the installation of the Scophony large screen television system there.[5] In addition I invented and arranged for the production of a sound level control device for the compensation of audience sound absorption—a useful and necessary component of the system as the absorption of sound by a human being is much greater than that by an empty seat. The success of this device was proven in Leicester Square and it was subsequently installed in several cinemas, in each proving to be highly satisfactory.

However, as already indicated, technical work in the sound and vision aspects of cinema formed only part of my duties at Odeon Theatres. I was also involved in a variety of interesting, semi-technical problems associated with the Odeon, Leicester Square. This theatre was to be superior or at least equal to the best and most luxurious cinema in England. As in all other Odeons the sale of sweets, chocolates, and ice cream was big business, particularly at the "Mickey Mouse Club" children's performance before the regular opening times of the theatre. Unfortunately, some children and perhaps also some adults used to smear the seats with ice cream and chocolate and it was not always possible to repair or exchange the dirty, sticky seats before the patrons for the following performance came in. The solution to this problem which was eventually adopted, and in which I was consulted, was to have all the seats covered with a yellowish material with irregular large patches like the pelt of a tiger. On such a pattern stains could not immediately be discerned under the dim lighting during the show. It was then to be the task of the usherettes to inspect all seats before the next performance for adults, putting a "Reserved" card on any dirty seats they found. After the last evening performance these seats were exchanged by the cleaners and maintenance staff who came in every night.

Another novelty I created for Odeon Theatres also had very little to do with the engineering division, but was not without its interest or relevance. In 1938 we learned that Odeon had acquired the rights to have *The Mikado* premiered in its cinemas. The film version of Gilbert and Sullivan's operetta contained most of the old, well-loved songs, and to be able to hear them repeated might well attract a larger audience. Thinking along these lines, I had an idea. I devised a method for the audience to order encores of any number, in the same way as was possible in live theatre or music hall. The manager of the cinema had to inform the audience of this facility and to ask them to applaud strongly if and when they desired to see and hear a particular scene again. The film would be stopped, rewound to the start of the applauded scene and replayed from there.

On the test run in a provincial cinema early in 1939 the innovation was well received, but by the time the film was scheduled for general release war had already broken out. At first all cinemas were closed; then the blackout came but although cinemas were allowed to open again after a few weeks they played to consistently half empty houses. All films showed under these conditions proved to be flops and *The Mikado* was no exception. So it was that some innovations were stifled for reasons that had nothing to do with the technology or the desires of the cinema management. It was nonetheless an innovation which was briefly a success, and one of which I was proud.

But perhaps the technical achievements of which I was most proud in my work for Odeon Theatres were my inventions of a continuous film projector and of a differential mirror drum.[6] I worked on them for some weeks and once I had reached a reasonable stage of development with both I discussed their possibilities with my employers. They were satisfied that both inventions were worth future development but these were rather outside the immediate expertise of a cinema chain. Odeon submitted them for examination to the manufacturers of their projectors, Thomson-Houston. The director there wrote me a very appreciative letter offering to develop the devices directly from my rough sketches, and at their own cost. I accepted. Construction of the equipment was quickly under way in the laboratories in Rugby. This was the state of affairs when the calamity which had threatened Europe for so long finally came to a head with the outbreak of war. That inevitably put an end to the project.

As might be expected, the outbreak of hostilities was to herald enormous and in some cases devastating changes, although precisely how these changes would occur and how they would affect individuals only dawned on us gradually. For me the first obvious change was that my involvement in direct show business came to an end. As the phoney war continued the head office of Odeon Theatres Limited was evacuated to the country, but I stayed on in London doing a variety of odd jobs in an atmosphere of growing confusion.

Since I had come to London in 1936, my private life had been increasingly dominated by the recognition of the scale of the Nazi threat and the persecution of the Jews. During the last two years before war actually broke out my sister and I made it our first duty to get permits for at least temporary refuge in Britain for those relatives and friends from Vienna for whom we could obtain either financial guarantees or special jobs acceptable to the Home Office. To get them out of German hands constituted an obvious, all-embracing duty so that the reorientation necessitated by the actual beginning of the war was not immediately clear. Things happened within my close family which upset my whole life: the prolonged illness and death of my father, the suicide of my sister, the absence of civil courage on the parts of my Odeon colleagues. In addition I had to face my own internment as an enemy alien—this automatically meant the end of my employment with Odeon Theatres. When, eventually, I was released from internment it was to an elderly and helpless mother in the midst of the daily bombing of London, and to no money and no job. I thus had two major challenges, in both of which technical expertise and inventiveness were my only assets: I had to earn a living and to fight the Germans. But how? Time was running out.

3: Printed Circuits: Origins and Early Applications

Even while I was interned I had decided not to buy my freedom by joining the Pioneer Corps. It was an unarmed auxiliary corps to which many refugees belonged as it was the only branch of the British army open to them. A number of non combative services for the army were assigned to the Pioneers. I was confident that I could do more for the war effort.

With this in mind, I managed to earn the minimum money my mother and I needed by doing odd jobs which took up about half the week; the rest of the week I worked on my inventions. I decided to return to the research on printed circuits for, although I had buried the idea in 1936 after the unsuccessful interview with Plessey, the war appeared to offer new opportunities.

Winston Churchill's rousing speeches on the radio highlighted the need for engineering expertise. The Battle of Britain had been won with the help of radar. Electronic devices of all kinds were becoming more and more important. Indeed they promised to be one of the most decisive factors in settling the outcome of the war. Every possible means of manufacturing them was organised, and tens of thousands of workers were mobilised to that end.[1] To produce ever-increasing quantities of these devices, and to produce them better and faster than the enemy—without having to call in more workers and machines so urgently wanted for other war work—constituted the main problems of the day. My idea of printed circuit production appeared to promise the solution to this problem. My thoughts thus returned to consider the whole concept of printed circuits.

Already in 1936 I had not limited my thoughts to the proposal of prefabricating the wiring of an electronic instrument, even though the example of the printed wireless set shown to Plessey might have appeared as such. Many methods have been conceived with that limited objective in view, and have dealt merely with the problem of metallizing a pattern on a slab of insulating material. However, the wiring, particularly in mass-produced equipment, is generally one of the least troublesome and least expensive operations. The currents

26

being very small, almost any metallic path will suffice from the standpoint of conductor quality. If demand were only for such equipment there were methods and means available which were very convenient and could be most easily improvised—such as silk screen or office-type printing using silver inks, or metal spraying through masks. It was my intention to introduce a far more comprehensive method for producing complete circuits.

For me the task of prefabrication in the electronics industry by printing methods was fascinating long before I had a chance to work on it full time. I was conscious well before the war of the importance and fast-growing scope of the electronics industry and the increasingly high standard of quality demanded. Electronics became ever more important in most types of communication and in information techniques from sound film to wireless telegraphy, and from control instruments in chemistry, biology, and electric power production to distribution, photo circuits, and temperature sensing.[2] There was no limit to its trend of expansion. In no other branch of the electrical industry were there so many different basic elements comprising conductors, semi-conductors, and insulators forming complicated networks.

My idea was that the printed circuit technique should be able to bring this network into existence as one integral structure, more or less simultaneously, in the manner of the inscriptions on a printed page or in the manner of a number of superimposed pages, as in a brochure. Even those elements which could not be printed by it, whether because they were irreducible to print or because their specifications were not fully understood, might at least be assembled into the rest of the printed network to some advantage.

The focusing of experiments and work on electronic circuits appeared to be a most fertile field for the development of printed circuit technology, despite the risk of overconcentration on only one section of the whole electrical industry. Indeed my early ideas had gone far beyond practical war applications and the mere prefabrication of the wiring. I had always realised that the printing of components could involve further development, using the expertise if not, the plant of the components industry. While it might be possible for a manufacturer of a component or for the assembler of a great number of identical electronic apparatus to choose a special production method best suited to this requirement, anyone specialising in printed circuit manufacture would face the problem of method selection from quite a different point of view. Such a method might not necessarily be tied to one process, but it needed at least to concentrate upon that one which would prove to be most widely applicable. What was required was a method which could be developed to produce almost all circuit elements theoretically printable, the results being of at least as good a quality as that of the conventional elements. Finally the products had to give the equipment manufacturer some advantage, for instance cost reduction, quality improvement, or both.

Obviously such a daring concept could not be developed in one go nor be introduced at any time, in one lump, even if all the necessary science and technology were understood and readily available. In wartime Britain, with restrictions of so many different kinds, the whole idea often seemed no more than a pipedream. But dreaming of this wonderful new concept served perhaps as an escape mechanism for me from the despair and helplessness I felt when rumours and news filtered through to me in London, concerning the fate of my relatives and Jewish friends who had not been able to escape to Britain or America and who had remained on the Continent at the mercy of the Nazis.

It was becoming more and more obvious that the situation was desperate for whole sections of the community. Awareness of this drove me ever more urgently to seek a practical realisation of my idea. I became convinced that printed circuitry could be the key to mass production of all kinds of electronic devices. I developed a technique which used existing and well-tried methods of the printing industry in a novel combination; I believed it would be useful in the production of electronic instruments and electrical equipment generally.

During the war printing was not considered by the Government to be an essential industry in the same sense as engineering, which could be readily converted to direct war production. Printing was necessary for maintaining morale, for help in administration and so on, but it did not directly produce arms and munitions. In the very rough grouping of goods into "guns" and "butter", its products decidedly belonged to the latter category. The printing industry was, therefore, not mobilised for the war effort in any but a very indirect way, and its output for civilian use was severely restricted. Consequently, a great part of the productive capacity of the printing industry lay idle. My idea of printed circuits aimed at putting to use the resources and well established practice of this industry and to satisfy the need for an almost unlimited supply of fast changing electronic components.

Although still confronted by the need to earn my livelihood I was elated by the conviction that now my opportunity had come to make a really important contribution to the defeat of the Germans. At the time I had no thought of monetary gain for my invention, although I was quite conscious of its wide reach. Many well-meaning friends and business men, who in later years learned of my invention, failed to understand why I did not look after the money side as well. Looking back, I do not fully understand why either, except that, while the research was unfolding my mind was on that and the possibilities it offered, regardless of financial reward. I was concerned principally to find a way of getting my ideas known and accepted. And I could not afford the time-delay for financial negociations. The German war machine became ever more threatening.

Although I was convinced that I held a trump card for the war effort, I did not go to the Ministry of Supply or to any large firm. Either move would, I felt, have meant burying the idea alive within the impenetrable layers of large, inefficient and lethargic bureaucracies. Instead I made enquiries as how best to proceed among my own friends and associates. My first important contact was with a Mr. White, the head of a relatively small but powerful financial organisation in the City. His organisation, I hoped, might provide at least the monetary means to develop the invention for patriotic reasons rather than for hope of gain. Sadly, however, I was not successful, even though he seemed convinced of the merits of the case.

The reason for my failure only became apparent later. While I was in the waiting room in front of his office, a young man came out. From pictures I saw after the war he must have been Frank Whittle, the officer who developed the jet engine, and whom White helped to a great extent. Many years later, when the part which both the jet engine and the printed circuit had played in the war was well known, I learned from White that he could not have supported both projects and so he had chose the jet engine. In retrospect I think he was right, although at the time, knowing nothing about Whittle and his jet, I was disappointed.[3]

After subsequently and also unsuccessfully, trying a number of small electronics firms I was finally recommended to a H. Veazey Strong, the head and owner of Henderson & Spalding. This was a very old firm of lithographic and music printers; I was told that even Beethoven's music was originally printed by them. Their plant had for many years been situated near a large gas works in the Old Kent Road in the Camberwell district of South London. During the Battle of Britain the Germans had tried to dive-bomb the nearby gas reservoirs but had instead hit the Henderson & Spalding plant, leaving it a complete ruin. As a result when I contacted Strong their only remaining activity was music engraving, carried out by a few old men not liable for military service. For this activity and the associated administrative work they had taken office space in London at 32 Shaftesbury Avenue, very close to Piccadilly Circus. It had been easy to do this because by then the cost of such central accomodation was very low, with most pre-war occupants having vacated it in the wake of the blitz.

Strong was a wealthy, enterprising, and liberal gentleman of the old type and a Master Printer himself. He apparently took a liking to me and to my modest requirements—a salary of £8.00 per week. He was particularly impressed by the invention, which would shift his printing output from the "butter" category to that of "guns." He was perhaps also able to see through it a way in which he could get back into printing during the war.

He accepted my proposal, and the only additional requests were that I would look at the instrument which the company has acquired before the war to overcome the growing shortage of music engravers. This shortage had existed even in peace time partly because of cost and partly resulting from union policy. To overcome it Henderson & Spalding had acquired a kind of music typewriter called a Technograph. This device, however, did not work. Its origin was apparently unknown and they had not found anyone who could get it to function at all; it was their hope that I might be able to help. We agreed further that applications for patents for my invention would be made in both our names. I was completely happy about this, and the one-man Instrument Division of Henderson & Spalding was born.

I was very much taken by the personality of Strong and felt that he deserved my absolute and unlimited confidence. I behaved as if he were my father and perhaps it can only be understood through this feeling that I did not even read, before blindly signing, the agreement between us in which I assigned to him all future patents on my invention for the nominal payment of one pound. We were in a taxi, with Strong's solicitor when Strong handed me the agreement for signature, introducing it as a formality. What mattered was only that I had the same confidence in him that he had in me. His solicitor asked whether I had a lawyer of my own who would like to see the document. I said I had no solicitor and he replied that he had assumed that, and had therefore taken care of both parties. I signed without any further ado and we went on to celebrate with a sherry.

Despite the suspicious implications of this episode when I read the agreement a few years later, my confidence in H. V. Strong was not perturbed. After all, I had achieved what I wanted most: to develop the invention so that it could be used to become a highly effective contribution to beating the Germans in the war. I had done that.

I thus started work with Henderson & Spalding in 1941 and at that time I envisaged my contribution to the war effort as pursuit of the best and most practical method to produce the widest possible range of electronic equipment by printed circuit technology. I also kept in mind that production could be carried out in wartime Britain where there was shortage of men and where the development of novel production machines was out of the question. Even before I started the thorough investigation to find the best method of carrying out my idea I naturally had to try to find out which type of electronic equipment was most urgently needed, and that for which printed circuit technology would be particularly advantageous or even indispensible. Asking the Ministry of Supply to give such information to an only recently released "enemy alien" was clearly inadvisable. There were, however, several vague ideas for guided weapon systems swimming around; gossip about them convinced me that some sort of electronic equipment in a shell might be the thing everybody was looking for. It would obviously need shockproof, miniature, and precise circuitry which could be produced in large quantities. I could not imagine any other production method for it except printed circuits. I was equally obvious to me that the designers had to learn at least the primitive features of printed circuit technology and that they had to be convinced that any "shell housed" equipment had to be based on that technology.

For my attempts to meet all these demands I was given the former caretaker's flat on the top floor of 32 Shaftesbury Avenue. This became my laboratory and was to remain so for nine years. I started by facing those principles on which the chosen technology had to be based.

In the first instance I considered the network of conductors in electronic equipment in the way the independent printer must see it. According to this point of view it is not enough to carry small currents when supported on a rigid base, or to use only ceramics or a similarly restricted selection of insulators, neglecting all the other substances which are necessary or desirable constituents of various other electrical elements. With certain forms of electronic circuits, or parts of such circuits, it may be possible to use second-rate conductors and insulators but no such tolerance exists for a great number of other circuits where the conductor must in every respect be as good as or even better than the wire already in use. I, therefore, chose as my first principle for method selection, that to be a universally applicable component the printed circuit must in all important aspects be at least as good as the best example of the electrical element or elements it replaces.

It should be noted that this principle of the equality or superiority of all printed circuit constituents is derived from the universality need of the printer who cannot hope to use the immense production capacity of a proper printing plant for the manufacture of merely one or two classes of special circuits, even if assured of a virtual monopoly. The universality aim and the quality standard derived from this aim are consequences of the inherent productivity of printing, and do not spring principally from the usual motifs or designers or other authors of specifications for a particular piece of equipment. The aim is usually an advance on the standard of the specifications. The quality postulate leads to a further general principle of method selection.

Since their introduction it has been argued that the very idea of printed circuits was a retrograde step. Modern industry believes in specification and the division of the labour process. The breaking down of any complex structure into many small and specialised units enables each to be mass produced under the most ideal conditions for the particular unit without need of a compromise between perhaps contradictory requirements of different units.[4] Printed circuit technique which aims at producing all or at least many of the different components of a circuit by one process seemed to deny this principle. If such were the case it might indeed have constituted a very doubtful progress. However, this was not part of my thinking. I realised that the "breaking down of any complex structure into many smaller units" necessitates their eventual reassembly into said complex structure. Avoiding such reassembly ought to be a first aim. The decisive question is therefore to find or create a technology which extends the automatic production of complex structures in comformity with the requirements of most of the small units. The balance of non-integrated units can subsequently be dealt with as at present.

In 1941, in the midst of the London blitz, I started an investigation of all the methods which I already knew or which I deemed to hold promise of being of use in printed circuit work. In the course of that investigation additional methods were found or invented and all techniques were subjected to extensive tests in an attempt to assess their potential value. As might be expected, no method secured full marks in all the various tests that were made. The crucial aspect of being nearly universally applicable and at the same time compatible with specialisation was not attained by any of these methods. Even assuming the complete development of each of the processes investigated it was not possible to see how any single one could produce all the various electrical elements to the same quality as contemporary component manufacturing methods already achieved. It seemed then that the printed circuit had only the limited scope of an assembly and constructional aid, that it was destined to become merely one of several means in the hands of set manufacturers to speed up the assembly line, instead of being a major revolution in the whole production of electrical and electronic equipment. This meant I had still not solved the problem I had set out with.

The breakthrough came when a whole new approach to circuit prefabrication was opened up by my invention and development of the technique based on the use of foil.[5] This technique proved to be the only one which passed the combined test of universal applicability and compatibility with the requirements of specialisation, and at the same time showed additional merits when compared with most other methods. It is, of course, not superior to every process in every respect, but the balance of advantage is weighted so heavily in its favour that all other processes seemed in comparison to be restricted in use.

It was always quite clear in my mind that the printed circuit was not just a clever, cheaper, or better method of making wireless sets or electronic wiring. My aim was that it would be a new technology applicable to nearly all branches of the electrical industry for the design and construction of solid electrical equipment from surface elements, as distinct from the conventional design and construction of this equipment from body elements. It therefore entailed the redesign of the three-dimensional structure of conventional electrical equipment and of more complex electronic equipment which at the time featured a mass of wire interconnecting distinct components such as resistors, capacitors, and coils. The design had to reduce this three-dimensional network into one, two, or more plane patterns of flat electrical conductors or semi-conductors, carried on one or more insulating bases in the form of sheets, films, papers or slabs.

In circuits comprising two plane patterns these patterns can be superimposed on the front and back of the same insulating base like ordinary print on the front and back of a page in a book. In circuits comprising more than two plane patterns they form a thin pile of superimposed and interconnected multilayers also like pages in a book. The superimposed layers, however, are laminated or otherwise fixed to each other, not just bound on one side, like the pages in a book.

The solution of the problem of combining the usually diametrically opposed requirements of universal applicability and specialisation by the foil technique is perhaps one of its greatest merit. The universality request springs originally from economic reasons while that of specialisation emerges from quality considerations. The successful combination achieved by the foil technique was to have further important consequences and its worldwide success is to a large degree due to this combination. The name "foil technique" came from the fact that the two-dimensional pattern of conductors which eventually was to replace the wires in all electronic equipment is produced out of an insulation-backed thin metal foil.

This insulation-backed metal foil became the raw material for the electrical circuit printer and is thus equivalent to the printing paper of the conventional printer. Just as the paper in all required varieties and exact specifications was produced in a special plant by specialist paper experts, so the metal foil was to be produced in special metallurgical works with complete freedom to use the most suitable methods and plant for its production, subject only to economic restrictions. With time these works were able to employ all available controls necessary for the single purpose of producing plain foil.

A further technical specialism to emerge from my early work was the backing of the foil by an insulator; this had to be done using a special converter and coating the foil with or bonding it to an insulator, itself produced by a specialist supplier. Top quality raw materials were needed and obtaining this meant relying on the integrity of the supplier, or wasting valuable time carrying out a series of tests. The prefabricated foil clad-insulation enabled the printer to check the quality of his materials prior to the printing process. He could test and select from the various qualities of any insulation backed foil before starting production or even design, and could adjust the circuit drawing to accomodate the rsults of the tests. With enough time and rigour then the final printed circuit should have no defect in any characteristic already tested and passed on the prefabricated raw material such as uniformity of dielectric, conductor thickness, tensile and compression strength, pliability, ductility, elasticity, grain orientation, temper, composition, resistivity, temperature coefficient, temperature endurance, and bond strength. To load the printer with the tasks of achieving even only the more important of the above characteristics would be quite impractical. It would not be too long before the printer could obtain his foil from specialist converters; foil which was finished, tested and controllable, with all characteristics known, ready for him to do his own specialist tasks; the printing and subsequent operations, most of them being processes already practised by the firms belonging to the industrial group "Printing and Allied Arts" or being processes well known in the electrical and finishing and converting trades. (Allied Arts are the arts used to produce a printing plate.) Such an invention, first developed in 1942 and still in unchallenged practice on such a large scale in 1988—and probably still in the foreseeable future—can claim to be based on correct methodological principles.

But in wartime Britain, when I was developing the basic technology, all these achievements were somewhere in the distant future.

The Printed Circuit

Until 1943 the components making up a circuit had to be connected together with wires.

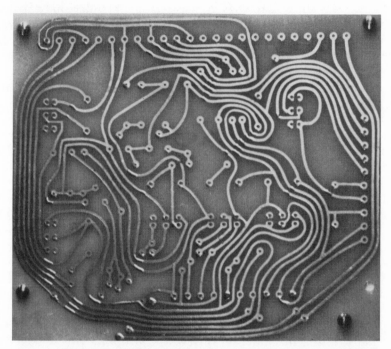

Simple printed circuit board. In 1943 the invention of the printed circuit board by Paul Eisler brought about a revolution in the electronics manufacturing industry. Wiring is replaced by ribbons of conducting material etched from a flat sheet covering the board. The flat ribbon pattern is printed onto the surface of the conducting material using an acid-resistant "ink". The unwanted conductor is then dissolved away in an acid bath, leaving behind only the desired ribbons. Hence, printed circuit.

Chip. The individual components on the surface of the chip are similarly connected by a pattern of conductors.

I might have been convinced of the revolutionary nature of my work, but I still had to convince others and to grind my way through the system. I was allowed to engage an assistant at Henderson & Spalding and took on an old friend of mine from Vienna, a self-made radio mechanic named Gustav Parker. He and I together made the first demonstration of the foil technique: a simple radio set demonstrating, without miniaturisation, this first step. The choice of a radio set as the first example for the demonstration of printed circuits in war-time London of 1942 was fairly self-evident. The processes of the foil technique needed for this demonstration were also more or less determined by the conditions in which I worked and by my conviction of the superiority and practicability of these processes, which would be the most practical within the foreseeable future at least for printed wiring and discrete components.

Indeed, when industry eventually took up printed circuits, years after the war, the most generally used raw material was the copper foil clad insulating board such as a phenol formaldehide paper based laminate to which a nominally 0.001" copper foil had been bonded under heat and pressure on one or both surfaces. In the London of 1942 such copper foil clad Bakelite boards were not available. All we could buy was rolled copper foil and we bonded this foil to varnished paper. This was our first raw material, our equivalent to the printing paper of the conventional printer and it permitted us to use the normal printing machine, an offset-litho proofing press, which together with some photo-mechanical equipment had survived the bombing of Henderson and Spalding's works in Camberwell.

There was obviously no opportunity at this stage to demonstrate the possibility of reducing circuits to the miniature sizes of postage stamps as the discrete components available at the time were anything but small enough for that purpose.

Throughout 1942 I had long discussions with O'Dell, senior partner of the patent agents Sefton, Jones, O'Dell & Stephens. I was guided by him in the applications I was to lodge, as far I could be. If my employers, Messrs Henderson & Spalding, had allowed me sufficient funds for patent applications I would have followed O'Dell's advice and lodged many applications at once. Having permission for only one application I struggled to make it at least as comprehensive as to indicate the extent of the application of my invention within the limits permitted by patent law.

When a single patent application was eventually lodged on 3 February 1943 it contained examples of the application of printed circuits not only to electronic equipment but also to the wide field of electrical engineering as it described and illustrated cables and interconnections, aerials, transformers, motors, and even valves and heated wallpaper. When this application was prosecuted it had to be divided and we selected three specific parts of the invention—the maximum we could do within the limits of the financial means I could persuade Henderson & Spalding to make available. One part dealt with the basic problem of the three-dimensional structure and printed circuit technology, the second with the actual manufacture of printed circuits by the foil technique, while the third contained some selected positive methods.[6] The latter was taken out essentially for the purpose of legal defence.

Allowing for wartime conditions in London, these patent specifications were as comprehensive as possible and still make interesting, readable documents. As far as patents, accompanied by research, development, and pioneering work, essentially by one man in a small laboratory and minimal plant can do, this work formed the origin of the electronic revolution which has affected all our lives since the Second World War.[7] It was a revolution which still influences our lives and will presumably continue to do so in the forseeable future. Some twenty eight years after the patents were filed, and thus long after they had expired, one of them had to be judged by the House of Lords. It had already been upheld in the High Court and the Court of Appeal, and was given the rare accolade of constituting the method which engineered the revolution of the electronics industry. This Judgment was arrived at in 1971. Early in 1943 I had one major problem: how could I communicate with the to me unknown designers of secret electronic equipment? I discussed this with Strong. He had a lot of connections and he undertook to invite to my little laboratory in Shaftesbury Avenue the various allied missions. We decided to demonstrate to them a simple but well known example of electronic equipment; a radio set, a much improved repeat of my 1936 presentation to Plessey, this time produced by the foil technique. I believed that if through my demonstrations I could make clear to the experts of the missions the principles of printed circuits they would convey my invention to their "back room boys". And so, eventually, it turned out.

From February 1943 onwards, hundreds of engineers and military personnel, British, American and Allied, attended demonstrations of printed circuits. While concentrating on methods for and examples of relatively simple circuits such as in a radio receiver easily applicable to wartime production, I pointed out my views on the immense possibilities of the new approach and they all came forward with queries, problems, and suggestions.

Depending on time available I explained to them and discussed with them the practical steps of the production in detail, just as I had set them out in my patent application. And while they were in the laboratory and could watch the proceedings Parker, my assistant made sample prints on copperfoil backed by a varnished paper, etched them with ferric chloride and cleaned them. They were then cut in small pieces which were distributed to the members of the mission. To the heads of the missions I gave a copy of the chassis of the radio set of which we had prepared twenty four.

At this stage many representations were also made to various large electronic firms and the appropriate ministries to make use of this idea in war production, even if only in an improvised way. However, the British Ministry of Supply rejected the invention for use in military equipment, and not a single industrial firm or Government department in this country could be found who would give the invention even a trial. The seed seemed to have fallen on barren ground. I was very disappointed. However, during the demonstrations in Shaftesbury Avenue the Americans had—unknown to me— picked up the idea and their National Bureau of Standards developed a proximity fuse using a printed circuit.

The proximity fuse is a miniature radio emitter and receiver fitted in an anti-aircraft shell (or to all sorts of missiles). It emits a radio wave during its flight and receives the reflection of that wave if a metallic body, which up in the air can only be an aircraft, is near that fly path. The received reflection sets off the explosive charge of the shell. While the anti-aircraft artillery so far had only impact—or time—fused shells which could rarely score a hit, the proximity fuse could bring down an aircraft which was within about 300 yards of its path. Thus the efficiency of the anti-aircraft fire was improved by many orders of magnitude.[8]

I had shown how printed circuits could be made by established printing methods quickly and in great quantities; they could be made as small as postage stamps and in multilayers, none of which would have unduly taxed Britain's wartime printing resources, nor its manpower. It is still a mystery to me why it was left to the Americans to take up my invention. Sir Henry Tizard, who was one of the high ranking advisors to the Government, and with whom we had had contact told me years later that he had tried hard. Perhaps we might have done better if we had tried to get an interview with Professor Lindeman, later Lord Cherwell, who had the ear of Winston Churchill.

The American scene was different, The National Bureau of Standards collaborated with the Central Lab. Division of Globe Union Inc. of Milwaukee, Wisconsin, and Lowell, Massachusetts, to fit their facilities. They adopted a much coarser, less miniaturised technique by printing silver paint on small ceramic plates. Initial production still reached not more than 5,000 a day, although they had no labour shortage. As a result only pre-production quantities were available to the allies in 1944. Nevertheless, the role which the proximity fuse in combination with radio location, later called RADAR, played in the war against Germany was for me an outstanding achievement, and went some way towards fulfilling my ambitions. About 4,000 of the V1's "Doodle-Bugs" aimed at London in a concentrated onslaught were destroyed over South East England by anti-aircraft shells fitted with proximity fuses. So many V1's could have caused terrific damage and casualties, and I was proud to have helped prevent yet more devastation. The other major success of the anti-aircraft artillery using proximity-fused shells was at the bridgehead at Antwerp, which was the principal entry point of the most vital supplies for the Allied invasion of the Continent of Europe. After the war I met General Armstrong in Paris. He was in charge of the anti-aircraft defence of Antwerp when the Germans attacked with a fleet of V1's. He told me that 97 percent of the hits by anti-aircraft artillery were with proximity-fused shells.

The confidence the Americans had in the proximity fuse is perhaps best proved by the report that millions of proximity fuses were distributed to and ready for operational use by the Pacific Fleet for the war against Japan in 1945. In the event they were not used as the Americans relied on another secret weapon: the atomic bomb. However, it was clear that the Americans took my invention seriously and realised its potential. In 1947 the story of the printed proximity fuse, which had formed an official secret until then, was released. After that release, printed circuits became established as an important branch of the armaments industry and in 1948 the U.S. authorities ruled that all electronic circuits for airborne instruments were to be printed. The publication of the story of the printed proximity fuse decided the fate of printed circuits. The release of the story brought a drastic change of attitude.

The deciding factor in turning the industry towards the adoption of printed circuits, however, was most probably the attitude adopted by the United States authorities. American military thinking was not only responsible for accepting the idea for use in the form of the proximity fuse; it also helped with propaganda and development contracts, work in Government institutions and agencies and, last, but not least, by clear statements of policy.

Bearing in mind the revolutionary impact of the new technique and the change in mental orientation which it demanded from designers of electrical equipment, fairly rapid progress was made in a number of governmental fields such as guided missiles of all types, and miniaturised walkie-talkies, in addition to a thousand and one instruments and gadgets used in aircraft and radar. That was, moreover, the stage prior to the "marriage of the technique" with semi-conductor production.

The penetration of the civilian market was closely tied up with the foil technique. The success which it has had and the great potentialities it opened up to practical development invite some speculation as to its function. So revolutionary an invention, which runs counter to many powerful interests could not have succeeded after years of obstruction had it not been firmly rooted in basic industrial needs.

4: Printed Circuits Take Off: The Beginning of Technograph Printed Circuits Limited

The inadequacy of financial rewards for what I was convinced constituted important technological achievements did not bother me much. I had already, in the agreement of 1941 with Strong, handed the decisions and the responsibility for the commercial and financial side to him. At that time, my aim was not to amass money but to help in beating the Germans. It might have made sense if after the war I had reorientated my mind towards money. But I did not do so. The fascination of trying successfully to solve what at first seemed difficult or insoluble technical problems kept up my belief in driving the foil technology to higher and higher goals. And I was in high spirits. I had fallen in love and got married. My wife was a dedicated scientist in her own right and as long as we could make ends meet we did not think of money.

At Henderson & Spalding the coming of peace brought a number of changes. While I continued with my work on printed circuits I also had to turn my attention to other technical aspects of the firm, as had been agreed when I originally joined them. Indeed immediately after VE day I had started to repair singlehanded the process camera, whirler, and other photomechanical equipment.

Commercial deals, including negotiations with other organisations about our developments and my inventions, were henceforth handled by a newly appointed sales director, a Major Holman. Before the war he had worked on the sales side of Henderson & Spalding, our mother firm, and had specialised in lithographic printing. As things turned out this was an unfortunate appointment; and indeed from the beginning I found him sometimes difficult to deal with.

However, I was not destined to do so for long at this stage, for the Instrument Department was to move. When I joined Henderson & Spalding this label applied to me alone, but in the course of the war and postwar years my department grew from one man to a small group engaged on and specialising in printed circuits. As a group we

41

were given part of the bomb-damaged printing works of Henderson & Spalding in Camberwell, repaired and equipped with some basic machines and devices to produce printed circuits. I was in charge as director and technical manager, and the laboratory in Shaftesbury Avenue was transferred as a complete unit, only the sales office remaining in central London. I was glad of the separation from Holman and the release from the obligation to explain to every visitor the rudiments of what we could offer him by the simplest use of printed circuits—a function which had been handed to me often daily and sometimes several times a day. The long journey to and from Camberwell was worth it for that separation and release. Besides my wife was at the time working in the Institute of Psychiatry at the Maudesley Hospital which was also in Camberwell. We could therefore drive a major part of the journey together. That made the change of workplace even more agreeable.

The move to Camberwell, and hence to more space, meant we were able to think about opening a factory to produce the items we were designing on a larger scale. But this would be an activity somewhat removed from Henderson & Spalding's main concerns in printing, so we in the Instrument Department remained separate. Eventually at the end of the 1940s the group and its subject were split off from Henderson & Spalding and formed into a separate company and given the name of Technograph Printed Circuits Limited.

When I started in 1941 "Technograph" was the name of the primitive typewriter for music which Strong wanted me to repair and remodel so as to become useable. By the end of the decade the "Technograph" music typewriter was quite forgotten but the name stuck. Strong organised a few friends to buy some shares which provided just the bare minimum for repairing the premises and installing the necessary equipment for production. And the opening of this first printed circuit factory came not a moment too early!

Staffing the factory was no problem. All my assistants who had worked with me in the laboratory at Shaftesbury Avenue came with me to the factory and carried on with the development and, for the first time, with production. Strong managed to get permission from the printers' union for us to engage a fully "learned" printer to run the offset-litho machine we had acquired and we engaged a few additional semi-skilled and unskilled personnel. Wages were the same as elsewhere in the printing industry and the atmosphere was that of a happy crowd inspired by the adventure of pioneering a new industry. I had little to do with the administration of the place which was in the hands of a highly qualified old friend, A. G. Lewis, our "Tony".

Henderson & Spalding's studio, which was housed in an adjoining

TECHNOGRAPH PRINTED CIRCUITS

32, SHAFTESBURY AVENUE, LONDON, W.1

Telephone GERRARD 4526

The Technograph Process (patented) can provide printed electrical circuits for almost any purpose.

Circuits may be in any preferred metal, combined with any chosen insulator, and of the many choices an outstanding combination is provided when the metal is bonded to phenolic laminate by polythene.

We shall be glad to discuss any problem with you.

We print:-

Inductances
Capacitors
Wiring
Fuses
Shielding
Servicing details

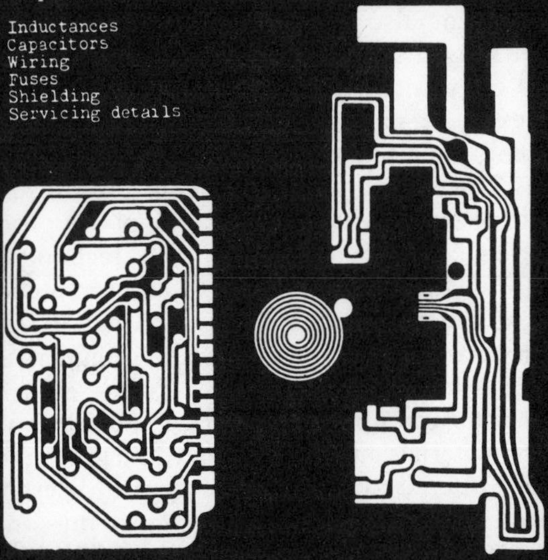

Post card advert produced by printed circuits process.

part of the rebuilt old premises, proved of great help. Our familiarity with the photographic equipment next door was of great advantage when we had to devise a method of improving the step and repeat composition of negatives for a very great number of miniature circuits (later named "micro" circuits) on one plate and render the whole process less costly. Strain Gauges, the size of which went down to about ¼ inch × ¼ inch, each one of which had to be inspected optically, were one of the first quantity productions requiring such plates.[1] We achieved success and a high production by controlling the step-and-repeat movements in the platemaking photographic process from the same mechanical guide as the corresponding movements in the optical inspection equipment of the finished gauges. This fairly obvious improvement also eliminated much of the test operator's fatigue by eliminating the need for optical registration on inspection.

In the work on printed circuit technology we moved steadily, generating and exploiting ideas which were to have some influence on later developments in electronics. In my book *The Technology of Printed Circuits* I described in ample detail the most directly successful and basic of the inventions and some of the products I could conceive and even launch while I was with Technograph. But the book did not go into the revolution of thought and concepts which the principles of the new technology imparted to the mind and constructive imagination of the scientists and designers in laboratory, drawing office, and factory.

Such influence showed up when the long and intensive research in semi-conducting materials and, particularly, the invention of the transistor in America, offered to postwar scientists and designers both the goals and the means of combining the new semi-conductor technology with the printed circuit principle which I had first conceived two decades earlier and which had become self-evident by then.[2] I am sure that, the basic concept of producing whole multilayer circuits by integrating passive components with the conducting network, instead—as previously only possible—of joining them to the network, must have contributed in the mind of the American scientists to the creation of the integrated circuits on a "chip".

My integrated multilayer circuits which, of course, could not in 1949 comprise such things as transistors, differ, however, in many respects from these later creations. Their combination of methods of integration by diffusion, the doping, the vacuum deposition, and other additive production techniques are perhaps the blunt reverse approach to my principle of separate, subtractive methods for each layer. I aimed at producing the whole extremely small circuits from original, pre-fabricated, pre-tested, multilayer materials and integrate them subsequently. In that way smaller quantities of "chips" as required for smaller markets could perhaps be produced economically as development may be quicker and simpler. The "software" problem may, in some cases, be eased. In spite of this difference of approach I believe it is quite probable that the ideas applied for in 1949[3] in Britain and their corresponding U.S. patent have contributed to the ideas which made the integrated circuits (IC's) of today such enormously important components. They are generally referred to as "chips". A chip means today a silicon wafer of a few millimeter square constituting a single package for hundred thousands of transistors integrated to form a micro-electronic circuit.

The great effort which, since the war, has been spent in America on research into semi-conductor material led to the invention of the transistor in the Bell Telephone Company's Laboratories in 1947.[4] It was a scientific breakthrough for the goal of the electronic switch. Of the three inventors—John Bardeen, Walter Brittain and William Shockley—Shockley got furthest. Still it took Jack Kilby of Texas Instruments till 1958 to get the first "simple" chip and thereafter it took many more years and the efforts of more firms to achieve the integration of huge number of transistors on a small silicon wafer.[5] In 1949 and in the UK these ideas were, of course, unknown or speculative, premature for the state of the industry, and could not be expected to bring in any money. Nevertheless, inventions cannot be valued only by the monetary reward they bring; speculative and premature ideas can be important promoters of progress and in this case patents are a good vehicle for them.

Needless to say, however, it was not all plain sailing for Technograph. The immediately postwar years may have been technically productive and interesting, but there were many problems arising mostly from our being a technical company with a board of directors whose expertise was in printing. Lack of experience undoubtedly led to mistakes when having to deal with outside institutions including large corporations, Government departments and the patent system. This all emerged when, with the move to Camberwell, came a wider general awareness of the possibilities of the new technology.

The British radio industry showed signs of waking up. Futuristic articles on computers and semi-conductors found eager readers and noises from America drew more attention. Perhaps there was something in that apparently crazy idea of printed circuits? Being the only technically qualified director on the board of Technograph, it was my duty to discuss the technical side of our business with potentially major customers or licensees and in this task I scored some apparently encouraging successes by getting our first licensees and first quantity orders.

I went to Cambridge for a long and searching discussion with Harmer, technical director of Pye—the makers of radio and associated equipment.[6] I won him over on all the points he raised, and when I felt almost sure of him I revealed the then not yet published development of the "Complete System of Component Integration", as well as my British Patent 690691, which I still considered as my highest achievement in the development of my original idea. Harmer understood and was visibly impressed: he even asked whether I could in principle "print" the still nebulous transistor as well. At that stage I could only be tentative in my reply as I did not yet know enough of that device to give a direct answer. Nevertheless I had clearly said enough: Pye became our licensee and Harmer joined the board of Technograph as Pye's representative. The agreement with Pye was negotiated between Strong's brother and Stanley, a director of Pye whom he knew personally. When I learnt its terms some months later I raised a stink about it in the board, as I found it so heavily balanced in Pye's favour. My criticism did not endear me to my colleagues on the Board and maybe marked the beginnings of wider differences between us to come. Harmer, however, had taken some interest in me, an interest which was to prove less than friendly. Not all our negotiations were surrounded by bad feeling, however.

The second licensee I acquired was the Telegraph Condenser Company in Acton, London. Sporing, their managing director, was a very intelligent engineer. From his accent I held him to be of working class origin looking at business from a superior, highly rational viewpoint. We understood each other at once and completely. He took a licence and started a small factory for printed circuit boards, after which point they required very little help from us. It was very rewarding to make this type of commercial and technical arrangement.

But such triumphs were sadly uncommon in our dealings with other organisations. I was still inventing and taking out patents, and it would have been completely beyond our means to manufacture everything at our factory: we had to make contact with others.[7]

The fate of some of the inventions (covered by the patents listed in the Appendix), which we developed at least to working models, is interesting not only for the technically orientated reader. It is revealing particularly for the insight it gave me into the interplay of forces and politics among executives in our large organisations.

For a start such organisations were always wary of outsiders and if they were interested at all it was with a determination to gain the best advantage for the organisation concerned, whether or not their interests coincided with the needs of the outside world. In addition the boards of directors could and often did simply choose to ignore an invention because it was N.I.H. (Not Invented Here). Yet these people wielded power, power to deny technological progress, or to push it in the direction of their interests. My experience in trying to deal with them made me even more sceptical of the use being made of patents.

The situation was not helped by the fact that among the directors of Technograph I was the only one who understood the technology—the experience of the other members of the Board was limited to the printing industry. Their way of dealing with the problems of the developing electronics industry, and the limitations on their approach may best be illustrated by the ways in which they tackled two extremely important aspects of Technograph's needs: the production of copperfoil by electrolysis and the manufacture of copperfoil clad board.

In the 1940s copperfoil—the basic raw material for our foil technique of printed circuit production—was only available in narrow widths produced by repeated rolling of thicker strips. The thinner the foil, the more work was involved in reducing the thickness of the strip and the more expensive the foil became. Moreover, no mill in England could produce wide copperfoil at all: there was no demand for wide copperfoil before printed circuits created the prospect of a market for it. It was quite clear to me and to my colleagues that the narrow, expensive, rolled copperfoil—the only type available—while good and cheap enough for the introduction of the new technique, had to be replaced. When printed circuits got into their stride a wide and basically cheaper copperfoil would be needed to serve as the standard raw material, the equivalent to the printing paper for the ordinary printer.

I could think of only two ways for the production of such foil: either by means of a Sendszimer or similar mill, or by electrodeposition. The cost of any mill of that class and the minimum quantity to run it economically were so fantastically high that this way had to be dropped immediately. Even though many years later when printed circuit production was already very large, an American firm got a Sendszimer mill and produced with it thin, wide copperfoil, at the time I emphatically favoured the other way: electrodeposition. Edison had already made ironfoil by electrodeposition; depositing copperfoil should be much easier and if the main production problems could be solved in the laboratory it should be cheap enough for us to afford.

I proceeded on that assumption and it proved correct. A young chemist on my staff, together with the mechanical engineer, rigged up a lead drum in a kitchen sink used as a tank. Within a few months they produced excellent copperfoil, down to a thickness of ⅛ of a thousandth of an inch, in a continuous length using a sightly adapted copper sulphate bath.

This success posed the problem of how to exploit it. It was out of the question that we ourselves should go into the commercial, largescale, production of copperfoil by electrodeposition. We had the knowhow, but what we had made could not be claimed as a new patentable invention. We had simply followed Edison. In view of the fact that demand for foil for printed circuits did not exist at that time, normal types of licensing agreement were out. Despite my optimism that the printed circuit had an important future it would not, I thought, have been easy to convince a big company that there was a potentially huge market for the product we were currently producing in the laboratory.

In the event, and almost by accident, however, we were successful in interesting a large firm in our development but we were quite unsuccessful in procuring any financial reward for Technograph. One day Goodfellow, an expert in the trade of certain nonferrous metals, visited us and when told of our development thought that the Royal Mint Refinery might be interested. We welcomed this thought and he introduced us to the Refinery. It belonged to Rothschilds, the famous bankers, who refined gold, silver, and other precious metals, and who were familiar with purifying such metals by electrolysis. The Refinery did not produce coins—that was the work of the Royal Mint, which belonged to the State.

We managed to convince the management of the Refinery of the future of printed circuits and, consequently, of the ultimately substantial size of the market for electrodeposited copperfoil. We showed them what we had done and, not being bothered by any financing obstacles, they started with the project in earnest and in complete secrecy. In a relatively short time the first production plant for wide foil was operative. Thus a credible source of supply for that raw material for printed circuits was secured. However, as to recognition for the proposal and help given, this famous organisation disappointed us. They invited us to glorious luncheons on the top floor of their premises overlooking the Tower of London, but they paid not a single penny to Technograph. Our directors, who had relied on a code of honour amongst respectable English firms, were naturally disappointed.

Initiating the production of wide, thin copperfoil as the most essential part of the raw material for the foil technique was the main requirement for supplying a hopefully budding printed circuit industry with the equivalent to what the printing paper is to the printing industry. But this was not the only need. As the first printed circuit radio set which we demonstrated in the early 1940s had already shown, the discrete components available at the time (and long after) required a stiff panel for their support with the network of their interconnections on one or both surfaces of the panel. The complete raw material for the main mass-products—radio and telephone sets—which we wanted to conquer for printed circuits, were therefore some sort of copperfoil clad boards. It was with the production of these boards that the directors of Technograph were to fail once again to obtain an advantageous arrangement for the company.

In our laboratory we had produced copperfoil clad boards for our own experimental and sample production using rolled copperfoil or our own electrodeposited copperfoil. It was not a simple job to bond the foil either to the available phenolic board, or to board made by ourselves with impregnated papers. We got a small, manually operated hydraulic press with electrically heated and water cooled platens and after prolonged trials and seeking the advice of several metallurgical and adhesive experts we succeeded in finding a satisfactory treatment of the copperfoil and of adhesive compounds ("bonding media") to give a very good bond not affected by the soldering temperature or by the other criteria we had stipulated. The bond could be produced under heat and pressure in the same press and simultaneously with the production of the board from the pile of the same impregnated papers as were normally used for the standard board.

We developed two types of bond, one for the etched pattern to remain proud of the board and one which permitted it to be pressed into the board and thus become flush with the board.

Finding the treatment of the copperfoil and the "best" bonding media by our somewhat enlightened trial and error development cannot, of course, rank as truly inventive but it was a necessary step and—allied to the production of the wide, thin copperfoil—completed the technical prerequisites for an industrial massproduction of copperfoil clad board, our desired raw material. Once again the question arose of how to exploit this development success. We, the minute Technograph organisation, could not, of course, dream of becoming a specialist plastic laminate manufacturer competing with giants like Formica or Bakelite.

A sensible move might have been to consult the Ministry of Supply, but we had good reasons for not wishing to do this. Since the middle of 1945 there had been a great commotion about the new technique of producing electronic instruments for the services among the higher civil servants of the Ministry, fuelled by or originating from head lines and cocktail parties with their American counterparts.

In February 1946 the American Bureau of Standards, Washington D.C., had released the first information on new techniques including many, by then published, on printing methods for production of conductor circuits. In 1947 a 43-page brochure on "Printed Circuit Techniques" edited by Dr Cledo Brunetti was published by the National Bureau of Standards,[8] prior to the publication of my patents on the Foil Techniques and other fundamental parts of the invention applied for on February 1943.[9]

The brochure dealt only with painting, spraying chemical deposition, vacuum processes, die-stamping, and dusting. Nevertheless there was such widespread interest by the scientific, industrial and American Government organisations that a general technical symposium attended by 700 people was held in October 1947 in Washington and the report on it was issued in a brochure in 1948: it seemed that a revolution in electronic production was brewing. However, even at that symposium, dealing with 26 methods of "reproducing a design upon a surface by any process," nobody had envisaged anything remotely similar to the foil technique. when my patents were published Dr Brunetti recognised at once their fundamental importance and came to see us full of praise: this was a moment of great satisfaction to me. The next issue of his brochure was amended to contain a chapter on the "Henderson & Spalding" method. This was most gratifying, but in England the reaction to what had transpired was much less forthcoming.[10]

Radio manufacturers could still sell all they could produce by the old fashioned wiring methods, regardless of price, to an eager public starved of new radio sets during the war and immediately afterwards. Such a seller's market was not receptive to new ideas or new methods of production.[11] Why should the managers of the British radio industry take up a new process as long as they made satisfactory profit with the old? We feared that it would take them even longer if they had no raw material readily available. Even if they tried, as was usual, to follow the American trend, they would seek knowhow from them not from us. That they might eventually get the knowhow from America instead of from us did not worry us particularly but the time delay did. Our first patents would soon be published and thereafter the whole situation would change in our favour, or so we believed.

With such thoughts in mind then we decided to approach, rather than to compete with, one of the big manufacturers of phenolic laminates, and we chose Formica. The British firm of Formica was a part of the De La Rue group with offices in Regent Street—a minute's walk from our place in Shaftesbury Avenue. We treated them as we had treated the Royal Mint Refinery; told them both what Dr Brunetti had revealed and our own story and expectations of printed circuits. We gave them samples. We shared with them our experience in bonding. We showed them our laboratory and the hydraulic press in the basement of Shaftesbury Avenue and demonstrated the bonding process on their phenolic board. They were convinced and promised to put the project on their development programme so as to be ready to quote, produce, and deliver when enquiries and orders were to come. There was no money or obligation to Technograph asked for or offered; from our point of view we were at least assured that a hold-up within the British radio industry for lack of raw material was unlikely. As in the case of the Royal Mint Refinery, technical expertise was made available by us to enable a large company to exploit what became, as we had predicted, a vast growth area. In both cases, the Royal Mint Refinery and Formica, the big organisations obtained the developments made by a weak, small firm free of charge. Factors of fairness or even recognition did not apply. Nor did the principle "Live and let live."[12]

The lessons we learned from these encounters were not, sadly, sufficient to arm us for later battles with large corporations and two subsequent examples of the struggles we had encountered show further how such organisations can operate.

The first of these involved our design of telephone exchange equipment, for which I had obtained a patent by the early 1950s.[13] The basis of the electromagnetic telephone exchange is a device, called the uniselector switch, of which there are a huge number, ranging into the tens of thousands, in each major exchange. The complexity of the wiring of the hundred fixed contacts of each switch with the interconnection of the tens of thousands of these switches in the smallest manageable space reaches the limit of the conventional art of wiring and occupies thousands of workers in the big manufacturing plants for automatic telephone exchange equipment.

From my student years in Vienna I already knew that the wiring in telephone exchanges was very complex and I knew also of the Siemens selector switch. Looking for an impressive application of printed circuits outside the radio industry it appeared to me that it merited an investigation. The most difficult task for me would be to prevent crosstalk. The simple means of twisting wires by which that was achieved was not available as a feature in printed circuits. The solution I found to the whole complex wiring problem and particularly the equivalent to the twisted wires was most invigorating to me. The first enthusiastic admirer was a certain Michel, who was in charge of development at the Nottingham firm of Ericsson, one of the small group of companies supplying the Post Office with telephone exchange equipment. I applied for a patent and had a simplified laboratory prototype made.

I vividly remember the outstandingly enthusiastic reception accorded to my idea and design by Gordon Radley, head of the Post Office. "That is the solution we have looked for, for a long time," he was reported to have exclaimed. With Ericsson strongly in favour and the Post Office so keen on it we felt victory in the air. On Michel's advice we offered a licence of manufacture to all the Post Office's big five suppliers (Ericsson, the General Electric Company, British Thompson Houston, Standard Telephone Cables, and Automatic Telephones & Equipment) on very easy terms and informed them of the subject in full detail in spite of the risks involved in such disclosure.

When the next meeting of the big five with the Post Office took place Radley went even further than we had ever expected and asked the big five to try the system out as an experiment in one exchange at the full expense of the Post Office. With such enthusiasm we were astonished to discover that this offer was refused by the Big Five. The reason given was that the electromagnetic telephone exchange was a thing of the past and no changes, whatever improvement they promised, should be made on and in it because the new "electronic exchange" would be ready to be installed in a very short time, in months rather than years. Nobody had any doubt of the prognosis that the electronic exchange was the type to come. But when? The timescale was the essential point.

Years had to pass before the Big Five's prognosis of the timescale could be proved to be quite wrong. Moreover, they had never made any pretence of its validity for the many hundreds of telephone exchanges abroad which imported the electromagnetic system of the British Post Office from this country. It seems that it was more convenient, and certainly less disruptive, to maintain the old system. The suppression of a major invention by an all-powerful cartel having no responsibility for the obvious damage it inflicts on the country, against the expressed policy of the authorised representative of the public, revealed forces of major magnitude capable of impeding progress; we could only have taken the matter up on a political level. We were never capable of that.[14]

Another line of initially promising business, where we were also unable to fight big organisations, concerned the technology I had developed for aircraft engine de-icing mats and de-icing mats for certain other aircraft surfaces. These were mats using only a small amount of energy. We designed heating mats comprising a printed circuit on a paper thin support for fixing them on the surfaces going to be iced up. When de-icing is desired an electric current is supplied to the foil conductors of the mats. They get sufficiently hot to melt their interface with the ice which thereby losses its grip and can be slided off.

In the early 1950 we made mats for the air-intake of the Rolls Royce Dart engine and the Viscount Supercharger as well as for the Armstrong Siddeley Double Mamba engine. I was also involved in producing the de-icing mats for the tailplane of the Bristol Britannia airliner which we made for the Bristol Airplane Company out of very thin nickel foil which we ourselves produced by electrodeposition.

These were our first achievements in the aircraft de-icing field which we thought of as a fairly rewarding business. We could offer excellent heat dissipation and weight saving and the mode of cyclic de-icing for large areas in slipstream. Moreover, the cost of our de-icing mats—while highly profitable for us—was so low in comparison with any other aero-engine or airframe component as to be considered negligible by the aircraft industry. However, our notable technical successes caused something of an uproar in adjacent quarters; we were dragged into a battle we could not expect to win.

It was an abrupt awakening. We received an invitation from Dunlop to visit them in their Midlands headquarters; Major Holman, our sales director, and I, the technical director, went full of expectations. After all, our famous great customers, Rolls Royce, Armstrong Siddeley, and the Bristol Airplane Company were very satisfied. So, after a guided tour of the Dunlop works and lunch, the meeting started; it all seemed very friendly, but the aim was certainly to get us to tell our story, in the course of which we revealed that we had no strong financial backing. Once this was established or perhaps con-

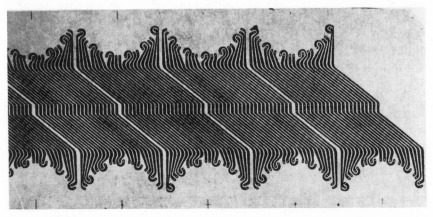

Printed circuit connections for telephone exchange switches, designed to avoid "cross-talk".

firmed to them, they started their attack. In a brutally coarse manner they made it clear that we must give up any attempt to get into the aero-engine business. That was their field. They did not need us and could punch the simple heating pattern themselves and they would punch us to smithereens if we did not at once stop interfering with them.

I had never experienced such a blunt and rude attack and I was shocked. I had hardly opened my mouth during the whole meeting as it essentially concerned business which was Major Holman's side. On the journey back in the train we talked about it and although my instinct was to fight the big beast I had to admit that all rational reasons were strongly in favour of giving up. Regretfully, we did that, as there was no other way for us.[15]

Our much admired de-icing mats for the airframe then suffered another blow. We had planned them as mats to be delivered to the airframe factory and storage depots wherever replacements or spares were housed, all over the world according to customers' wishes. They were readily mounted on to the airframe and basically cheap. The snag we came up against was the need to have any component for fixing to an airframe passed by no less than three commissions before it could be established as "airworthy" and be used. The bureaucracy involved in these commissions with their various establishments and the apparatus which an applicant requires to deal with is not only cumbersome and time consuming but also so costly that a small organisation like Technograph could never afford it. We did not want to put such a cost on what—for a printer—were only small numbers of mats. The aeroplane companies, on the other hand, refused absolutely to pass any such bought-in components through the commissions at their cost by their own staff. No exception could be made for us.

While we were, in vain, looking for a way out of the dilemma, a competitor arose who solved that problem by using an obsolete, and terribly expensive method of providing a de-icing pattern on an aircraft wing. Making the job so costly provided enough income to afford staff and the werewithal to deal with the three commissions. The customer seemed quite prepared to accept that this de-icing pattern would be very expensive, and did not raise much difficulty over the price.

What the organisation did was to hire a large hangar in Luton, near London, in which they metal-sprayed the aircraft wing directly with the de-icing pattern, fixed the necessary terminals and covered it with paint. We believed the foil technique to be superior technically, and it would definitely have been cheaper. The main expense with the choice of the metal spray was obviously the fact that it had to be done directly on the aircraft wing. The wing had therefore to be transported from the airplane company to Luton and all the way back, whatever the distance. Such transport was a very costly affair and the high bill presented to the airplane company was consequently understandable. Needless to say for most of the aircraft the taxpayer or customer had to pay eventually openly or indirectly.

I find this case so noteworthy because it shows that in our competitive society providing something better and more cheaply, something appreciated by the customer as such, is not the unfailing road to success. The very existence of a large bureaucracy and the cost of complying with regulations proved to be great obstacles to the progress we were pursuing.

By the mid 1950s, then, Technograph had had a number of skirmishes with bureaucratic organisations during its short history. But it remained a relatively small and powerless outfit; this meant that we were still not properly equipped to deal with what would turn out to be the biggest bureaucratic threat of all: the arrival on the scene of the National Research and Development Corporation.

5: Adventures Abroad

While much of our time during the decade following the Second World War was taken up with the domestic situation we were still eager to explore possibilities overseas. The results varied from the disastrous to the ridiculous but all contributed to the situation the firm found itself in by the mid 1950s.

The most obvious country to show interest was, as expected, the United States, as my ideas had already been taken seriously there at least by the military. But in spite of the great interest shown immediately after the war by the technical publications and despite the continued support of the new technology by the military establishments in America, the radio industry everywhere stuck to its point to point wiring methods. It was only after 1948, when the American authorities actually decided that airborne electronic equipment had to be the lighter and more reliable printed circuit type, that the American industry started to take notice. The growing awareness by the industry of the possible availability of transistors in the near future and of other new technologies—such as the development of computers—naturally reinforced their attention, which extended even to us in far away London.

One day in 1949 we received a telephone call from the Board of Trade introducing two American gentlemen and asking us to receive them. The first was William Garduner who imported certain electrical goods, such as the well known Garrard gramophone motors, from Britain to America. The other was Hubert L. Shortt who had a factory near New York, manufacturing certain electronic components for "big" American radio set manufacturers. Shortt's main line was a type of aerial consisting of spiral coils stamped onto an insulating sheet by the then heavily advertised Franklin process. These coils were similar to the printed aerials which formed the rear walls of the radio sets I had produced in 1942 for demonstrating the idea of printed circuits. As I found out later, during my first visit to America, Shortt had been strongly motivated to come to England for two other reasons linked to his own business. Firstly the development of new magnetic materials had started to render his aerial coils obsolete and threatened to bring about a shutdown of his factory unless he could

56

introduce a new line. Secondly, a director of General Electric—a major customer of Shortt—was a great supporter of his. He was perhaps even a partner of Shortt in his business, apparently with the full knowledge of the Board of General Electric. It was this connection which was the main source of Shortt's strength.

In view of the decline of, and anxiety over, the aerial coil business, Shortt and Garduner had, (quite probably with the director of General Electric,) hatched a scheme to become a supplier of the new printed circuit boards, (to General Electric in the first instance,) for new radio set models. Shortt was confident that he could change over his factory quickly and his friend was equally confident that he could help Shortt procure the order from GE. It was, after all, a well known policy of big corporations to let small companies try out new methods or products, to let them commit the initial errors and overcome the difficulties inherent in starting something new. The big corporation would take over when these risks had been minimised and a clearcut profitable run was assured. If the small fry had served them well, they would reward them well.

All this is, of course, mainly speculation about the background to Shortt's decision to enter the printed circuit scene. For the origin of the agreement we must go back to the Technograph scene in London 1949.

The atmosphere of this scene was conditioned by two features: a strong confidence in the imminence of the breakthrough, in other words, the start of the acceptance of printed circuits by the radio industry; and very little—if any—money in the Technograph kitty. I was essentially carried away by the first, but the brothers Strong were understandably very worried by the second feature, so much so that when my invention was already an acclaimed and proven practical success, they still felt as though I had led them up the garden path. They were obviously waiting for the Messiah and that wait was nervewracking. Perhaps Shortt took on the role of Messiah in their dreams when he appeared on the scene disclosing that—while he was a small manufacturer himself—the great General Electric was behind him.

He was at first turned over to me, and I made him familiar on a technical level with my inventions, my patents, and my work since 1941. He already had some idea of the main process and after a few sessions at which I also showed him round our small laboratory in Shaftesbury Avenue, he appeared a completely convinced enthusiast.

Then the Strongs took over. They had found out that Schortt's and Garduner's tale about the big corporations entering a new technical arena via a small pioneering company, officially at arm's length from them, was a well known American phenomenon. That was about all I learned from them prior to being told just before Shortt's and Garduner's departure that a verbal agreement had been arrived at by the Strongs and a draft agreement had been worked out in which Shortt would form a Technograph company to be the exclusive licensee for the United States. He could produce as well as sublicence others in the States. We would have 25 per cent of the shares in the American Company and a part of the royalties. I would have to go the States the following year for a month or so to help Shortt with the start of the production, by which time the agreement would have been finalised, Shortt would have transformed his factory, and we here would have moved into a new Henderson & Spalding factory at Sylvan Grove, Camberwell, as planned.

A big party was thrown at the luxury flat of Strong's brother in Kensington as he became confident of being able to raise money for us on the strength of the American connection. After all, to have a sizeable participation in an American company backed—at first unofficially but eventually—by one of the greatest American corporations should be attractive enough for some friends of his to acquire shares in the British company. That indeed happened, and provided the finance for the adaptation of the war-damaged section of the Henderson & Spalding works into our first new factory, for the basic equipment and a modest amount of working capital. The next move and the start of the factory kept me very busy and I never enquired about the final agreements.

It was a long flight to New York in 1950. I remember having a sleeping bunk and that we landed for refuelling in Newfoundland during the night. Shortt collected me and from that first moment, until I left America, he acted as tourist guide as well as commentator on the American industrial scene. It was only when I arrived in America that I learned that Shortt's friend at General Electric had been thrown off the Board. The reasons for this dismissal were not connected with Shortt. However, with the director's departure went the whole General Electric plan and their interest in printed circuits became obsolete. With this loss, the financial means which Shortt, Garduner, and their friend Spinrad could raise became very limited. They therefore gave up the idea of manufacturing printed circuits by Technograph Printed Electronics Inc., the company which they had formed and which was to have been the exclusive licensee of the British Technograph Printed Circuits Ltd., and thus the exclusive licensees of all my American patents. Instead they wanted to sublicence other manufacturers, a policy which in view of the already very impressive interest shown by industry seemed both to be very easy and to promise quick financial returns. Shortt and Garduner knew all my American patent applications which by 1950 were already an impressive number.[1] By application we had already secured the priority dates for the patents and I could talk about them and show examples, although for most of them I could not yet give the final wording of claims allowed.

In America it takes a long time before any application passes through all the tribulations of prosecution and is granted and published. My printed circuit patents were no exception to this delay as is clear from the late dates of issue. A great consolation for the patentee, however, is the fact that this delay—contrary to the law in Britain and in Europe—does not in the United States shorten the time in which the patent rights are valid. The life of an American patent starts from the date of issue and there are no renewal fees. Furthermore, patent applications are very thoroughly investigated, so that a granted American patent conveys a certain reassurance of valid rights.[2]

The role Shortt had assigned to me was that of the great pioneer of the miraculous process, from the strange distant country of ancient glory, someone who could solve all problems and knew all the answers. He had arranged interviews for me with the top engineers of many of the famous corporations. All these engineers already knew sufficient about printed circuits: they did not require general introduction. They had all given thought to the new technology, and what they required usually was the reasons why the foil technique was superior to and for them more profitable than any of the other methods described in Dr. Brunetti's brochure published by the National Bureau of Standards and made familiar to them by articles, press advertisements, and other means of American publicity.[3] They also sometimes had particular technical queries requiring serious discussion and experimental verification, which would only really have been possible if proper laboratory facilities had been available. At their base in Tarrytown only the barest minimum of equipment existed in a corner of the huge empty laboratory, hardly enough for making primitive samples. Although requirements for experiments occurred seldom enough, I did not like having to confess the need to wait until my return to London before being able to carryout a thorough experimental investigation of some particular queries. However, in spite of such occasional imperfections, I enjoyed my role as a kind of sales engineer for my invention and some of the firms actually became licensees of the American company after it had revised its policy.

I had expected to be back in London within a month but this first trip to America extended to more than two months. This length of stay—due essentially to Shortt's and Garduner's arrangements for interviews with firms, the Signal Corps and Air Force establishments, with our patent attorneys and their own advisers—also permitted me to learn a little about the state of developments in allied arts which I felt promised to influence strongly the further development of my inventions. I visited, for instance, the Eastman Kodak laboratories in Rochester where I learned about the state of development of the Kodak Photoresist which I had hoped would become available in London some time earlier. The Kodak Photoresist is a chemical compound (solid or liquid) for a uniform lightsensitive coating of the foil. (When and where exposed through a photographic negative to ultra-violet light it hardens and becomes resistant to the etching or plating baths used. On the unlit areas it is dissolved by the development solution. There the blanc foil can be etched off its insulating base.)

Then in Washington I was received with accentuated expressions of respect at the Patent Office by the examiner dealing with printed circuits. He showed me a cupboard, he had named after me, full of applications he had refused on the basis of my prior 1943 application. We clarified a few points regarding pending applications and I felt very satisfied with the progress made by two hours of conversation on issues which several years of correspondence had failed to clarify.

Although I was not directly involved in negotiations with prospective licensees I learned enough about some of them through conversations with Shortt and Garduner to become disturbed, with great fears of the consequences. Some of the negotiations seemed reasonable enough and did in due course lead to licences. In connection with others, however, such as with the computer companies, terms were insisted upon by Shortt relating the royalties not to the price of the printed circuit itself but to a very small percentage of the price of the whole computer! Such a licensing policy, given the inescapable need of computer manufacturers to use our technique had to be rejected by us. But this need of rejection induced so much fury in the industry that open infringement of the basic process patent started. It took many years before the American Technograph Company had amassed sufficient money from the royalties and other fees of its licensees to be able to challenge the infringements in court. By then the infringers had become legion and, aided by the absence of defence, had infused into the industry the belief that my process patent was invalid.

When I left America I had, of course, no premonition of what was to follow from this policy in later years, but there was already a cloud which I felt hanging not only over the American Technograph organisation but one which might cast its shadow over the British one as well. This I was sure would result from the pre-eminence of American industrial policies; Britain and Europe would only follow American tendencies, forward or backward, even when it came to my own technology![4]

Negotiations with the United States were not the limit of our experiences overseas; in many ways our attempts to enter European markets were even more frustrating, and in one particular encounter quite bizarre. Our introduction to the Continent came through one man in particular, Ben Smith, whom I first met whilst I was in the midst of getting started with the new factory in Camberwell.

I received a call from Harold Vezey Strong asking me to see Ben Smith, an American millionaire, at the Ritz Hotel. I was told that he was a big operator in minerals, including oil, and engaged in searches all over the world not only for sites and concessions but also for major inventions in industry. At the Ritz I met a sharp eyed, stout man I estimated to be in his middle sixties, obviously clever, very quick on the uptake and in decision, and of civil manners. After a few questions and answers concerning my personal history and position in Technograph he went straight into *medias res* and I had to give him in ten minutes or so the basis of my philosophy and of my idea of the future of the technology on which I had worked actively for the previous nine years. He appeared very interested and enquired whether he could come in with a large sum; he also wanted to know our position over American rights.

When I told him that we had already granted exclusive rights to an American group he cooled off. But then he moved his attention to France; my information that we had done nothing about it in that country, led him to say that he would put us in touch with his French agent. I got the impression that he took an immediate liking to me and that he only regretted being too late to win the American rights. Although we therefore did no business with him, he invited my wife and me to parties several times and I remember him as a basically lonely man in spite of his many daring deals and the swirl of activity around him.

Ben Smith kept his promise regarding his French agent and soon after my first meeting at the Ritz my wife and I received an invitation to visit Paris from a M. Rosen. My wife and I had been in Paris before, but this time we were given V.I.P. treatment with, for us, extraordinary luxury, French gourmandery and Parisienne entertainment fashionable at that time. Rosen had an office at the Place de Théâtre Français employing a few draughtsmen designing applications for a rubber bonding process for which he was looking for major licensees. During the war he had been in America; from there came his fluent English, and his idea of big business through the introduction of modern industrial developments into a technically backward France, which was still suffering from the wartime occupation on top of prewar neglect. He had been very successful in introducing the heating cable Pyrotenax to France and was, in view of Ben Smith's recommendation, keen to try his luck with printed circuits, a complete novelty for France. Anything which required understanding of technical matters he delegated to a young engineer named Charles Schwanhard who, although he spoke only French, making communication a little difficult, had an extraordinary intelligence. He instantly grasped any technical concept, was loyal, incorruptible and sincere. He was very cynical about politics and business and critical of everybody, including us and himself. We took an instant liking to him and became good friends.

Contrary to the state of affairs in Britain and America, Technograph had practically no basic patent position on the Continent. During the war the Continent was occupied by the enemy and after the war we had hardly enough money for patent applications in Britain and America and could not consider expenses for patents in these ruined states. What we therefore had to sell in France was knowhow for an industry which did not exist and about which only some technical journals had occasional news. It proved an uphill task in the course of which I had to go to Paris very often, sometimes weekly. I flew there in the early morning and took one of the last planes back to London.

It soon proved impossible to find personalities or industry really enterprising enough for anything we could suggest. Rosen, however, did not give up. Of all the connections he tried, the ones he pursued with the greatest effort were those with French governmental departments concerned with defence. Through them one episode happened of which I still have most vivid memories.

It was a most dramatic scheme, the real reason for which I actually learnt only months later: it was to use my invention to bring back to France a huge amount of money in Indo-China. I do not know what my reaction would have been had I known anything of the purpose of the whole exercise beforehand. In retrospect I can put together the fantastic story, but at the time it happened I was an ignorant cog, within a far greater machine, simply doing the job I was asked to do and enjoying the Riviera as if we were on holiday.

Nevertheless looking back it is fascinating to contemplate how enough money was raised—for an essentially crazy idea—initially to found a company in Monte-Carlo, the "Technelec Société Anonyme

Monégasque." Such a company would escape many of the French tax burdens if it made a lot of profit; it would also no doubt mask the whole scheme from the naive Technograph people in London and give an anonymous mantle to the negotiators in Paris.

In the capital, Rosen's governmental contacts had introduced him to a captain in the French army who had served in Indo-China. There he had married a rich governor's daughter and returned to France, where the couple had a beautiful villa on the coast about twenty minutes drive west of Nice. Their problem was how to bring their money, in value over a million pounds sterling from Indo-China to France given that the removal of capital or the transfer of money for any reason other than what the French Republic directly decreed, was prohibited. They had, however, discovered a loophole: if the money was needed for arms, permission might be obtainable. Once the money was in France, once this hurdle was overcome, the problem would be much, much easier for them.

I know that Rosen had tried for many months to get governmental funds for printed circuit production in France by emphasizing the importance of printed electronics for weapons, as the proximity fuse had demonstrated. He argued that a printed circuit production facility was of great importance, not only for the growing French radio industry but also was essential particularly for a modernised rearmament programme.

That line of argument could easily have been picked up by our ex-Indo-China captain, who was connected with the same officials, and from these beginnings the idea probably grew first to establish a small laboratory-type production unit near Nice; Rosen undertook to get Technograph's official licence, and my initial assistance, and the captain and a friendly Monte Carlo lawyer were confident of getting permission to invest the Indo-Chinese funds in this way.

The money to start up the scheme in an elegant way was apparently no problem. The first, and up to then, only thing we in London learnt about it, was a request from the Monégasque company for a licence, and an invitation to Mr and Mrs Talbot Vezey Strong (Harold's brother) and to my wife and myself to visit Nice with a stay at the Hotel Ruhl for a fortnight, all expenses paid. I was to get a new small factory going and Strong to negotiate and complete a licence agreement. It sounded like manna from heaven. Had Rosen found a millionaire benefactor? If it was a breakthrough, it was in style! We naturally accepted and set out to spend a fortnight in that wonderland.

The Ruhl was an old-fashioned luxury hotel, the Riviera was not yet as spoilt as in later decades and we had enough time and opportunity to live it up. My wife and I had never been in such exuberant French company before and we never asked why the mundane jobs of installing a small laboratory, called a factory, and concluding a licence agreement, were the occasion for such jollity and expenditure. We concluded that we would never understand the French and left it at that. When we departed primitive samples could be produced in Nice and Strong brought home a new licence agreement. And that was more or less that.

I learned part of the story from Schwanhard several months later and part I pieced together from disparate memories in the above account. Whether or how much of the Indo-Chinese money eventually flowed back to France I do not know; at any rate none came our way. The unique application of my invention by the ingenious French captain to the shady transfer of money to France, a field of application which was certainly outside my wildest dreams of its development, remains in my memory for the picturesque fun it engendered.

However, it would not be just to M. Rosen to forget, because of it, the serious attempts he had made to interest some large organisations in printed circuits. Of the large French firms to which Rosen introduced me the most interesting was the "Société Alsacienne des Constructions Mécaniques." I had several sessions with the head of this vast, rich organisation, a M. Boumelard—a highly educated elderly gentleman of the type you see in films portraying the soigné old-fashioned superior managing director. We got on very well but discussions dragged on and on and Technograph was kept waiting for a decision. Then one day I received a telephone request to meet Boumelard for breakfast at the Savoy in London.

He proved charming as always and said he had come especially to see me personally. He could get the order for developing and producing torpedoes for the French navy and the fleets of some other NATO countries, at least 2000 annually. Development was to be carried out in a laboratory on Lake Annecy. The provision was for a staff of 24 and he asked me to take charge of this laboratory. I could write out my salary and conditions as I wanted and he would give me a blank cheque. He would take care that I had all the equipment and support I needed from outside.

I had never had such an offer. Was this the great price won in life's lottery? Whatever it was, I was quick with my decision and told Boumelard at once that I could not desert Mr Strong, whose salvation, together with that of Technograph depended at that time on my remaining with him at least until 1956 under a service agreement. Boumelard expressed appreciation of my reason and we parted with regret. On my next visit to Paris I went to see him at his office, to keep in touch; but he was not there and I was told he was ill. I enquired several times in the following months about him until I learnt that he had died of cancer. I saw his successor once, but it was only to express my condolences.

Rosen's efforts on our behalf were not limited to trying to make good contacts in France. He also took me to Eindhoven in Holland where I spent a day with several engineers and scientists in Philip's laboratory; and he made an appointment for me with Telefunken of Hanover, Germany. We had failed to apply for patents in either Holland or Germany. In Germany my patents would have had to be granted and would have become unassailable if we had launched patent applications during the period when the Allied Occupation Status was valid. But we missed that date and after the publication of our process, no patent could be successfully applied for it anymore. As we could not offer a licence under patents, we got nowhere with either the Germans or the Dutch.

The only personal satisfaction I derived was from a lecture I was invited to give in German to the German Engineering Society (V.D.E.) a few weeks later in Hanover. After describing the process and some of its applications I showed a film of English anti-aircraft artillery shooting down German aircraft and doodlebugs (V-I's) by means of proximity fused shells. I left my audience in no doubt as to my attitude to Nazi Germany and emphasised that the prime purpose of my inventing and pioneering printed circuits had been to fight the Germans in the war. Only two people applauded. I may have been tactless and politically insensitive of me to remind a German audience of the facts but I still do not regret it.

One German contact and subsequently my very good friend whose acquaintance I owe to our efforts in Paris was Werner von Haag. He had been a high-ranking German officer and diplomat, somehow belonging to the circle of German officers who tried and failed to assassinate Hitler in the attempted putsch of 1944. He lived in Paris, was a reliable anti-Nazi and had fantastic connections. He, no doubt, convinced several personalities in Germany to take printed circuits seriously but in the event the German industry took licences from American firms. For, the industry in the United States had not wasted any time; it paraded the new technology in the press and spotlighted American progress. There was no liaison with Britain, which had lagged behind in its acceptance of this British-born achievement.

Overall, then, although my travelling salesmanship for printed circuits in France and other continental countries was not imme-diately financially rewarding for Technograph, it certainly contributed to the spread and eventual acceptance of printed circuit technology. I personally also became known as its inventor and pioneer in France. A consequence of this recognition in France was the honour I received in 1957, out of the blue, when I was made an "Officier de l'Ordre du Mérite pour la Recherche et l'Invention." I was also personally honoured by the Italians some time later. They appointed me "Academico Corrispondente of the Accademia Tiberina" in Rome, and accorded me other honours. True, it did not cost them any money but nothing of that sort ever happened elsewhere. It is, perhaps, ironic that my name is more clearly associated with printed circuit technology in Europe than it is in Britain, where I undertook the crucial development work of my invention— an invention which has become the basic production method of the whole electronics industry all over the world, giving work to many tens of thousands of people.

List of Licensees of Technograph Inc.

Note the names of some huge corporations known all over the world which Shortt signed up, although he had practically nothing to offer apart from the rights under my patents.

Alvic Products Company, Solon (Cleveland) Ohio.
Amerace Corporation, Gavitt Wire & Cable Division, Brookfield, Mass.
Ampex Corporation, Redwood City, Calif.
Baldwin-Lima-Hamilton Corp., Philadelphia, Pennsylvania.
The Bunker-Ramo Corp., New York.
Collins Radio Company, Cedar Rapids, Iowa.
Eastman Kodak Company, Rochester, New York.
Electronic Films, Inc. (Subsidiary—Xerox Corp.) Burlington, Mass.
Formica Corporation, Cincinnati, Ohio.
General Motors Corp., Detroit, Michigan.
International Business Machines, New York.
The Mica Corporation, Culver City, Calif.
North American Aviation, Inc., Los Angeles, Calif.
Photocircuits Corporation, Glen Cove, New York.
Texas Instruments, Inc., Dallas, Texas.
Western Electric Company, New York.

6: Enter the NRDC: The Kiss of Death

From the very early 1950s interest in printed circuit board manufacture rose steadily in America and even in Europe. We were very busy in the factory and it became clear to us that we needed to have a much wider financial basis than the meagre finances on which we had managed so far. We had two choices: the City of London or—in view of the national importance we could understandably and justifiably claim—the National Research and Development Corporation, (NRDC) at the time under Lord Halsbury.[1] We voted for the latter in the belief that the NRDC would leave us more freedom. It proved a major mistake.

The negotiations, not unexpectedly, took a very long time. I took no part in them, but even before the negotiations started in earnest and, of course, during the time of these negotiations we all had to be very careful not to take any steps of possibly major importance without first consulting the NRDC or obtaining their consent. Strong's bargaining position was poor. After the war he had tried to revitalise Henderson & Spalding using the old salvaged prewar rotary offset-litho machines; but he failed in this attempt and the plant had had to be auctioned off. He had lost his participation in the music publishers, Novellos, and the only thing left to him was Technograph. This put him in the position of needing the NRDC more, perhaps, than it needed him, and the NRDC was well aware of this. I had to stand by him in this situation, whatever I already thought—by that time—of his original agreement with me ten years previously.

The terms offered by the NRDC were uncompromising. Their support was in the form of a loan to Technograph and was dependent on various conditions which I considered most onerous and which proved to be so.[2] By the time the lawyers had completed their job and the contract was formally ready for signature, printed circuits were already becoming a booming business. The NRDC could not boast anymore of their daring to take a new technology "off the ground". Our factory was proving a profitable business, although not yet capable of paying back past deficits. The loan from the NRDC to Technograph could be looked at by this time as an investment in an already running business which wished to and was able to expand.

67

Expansion of production and continuation of development work were not only obvious needs for this country, they were also essential—given growing interest abroad, particularly in America and France—if we were to keep our leading position.

For, although my numerous trips abroad had not brought much business back to Technograph, they had demonstrated to me that the technology of printed circuits was spreading. But none of this had any impact on our negotiations with the NRDC.

The final contract provided for the NRDC to appoint two representatives on the Board of Technograph, each with rights of control like those of a managing director. Without the approval of these NRDC directors chosen by and responsible to the NRDC, no decision could be made by the Board. The other directors were left essentially only a kind of advisory function and the responsibility for carrying out the directives of the NRDC.

Our situation at Technograph was not helped by the two appointments made by the NRDC. One of their representatives was a retired brigadier from the Indian service. After Indian independence British personnel streamed back to the old country where the Government had to find employment for them. The brigadier sent to us was one of the NRDC's allocation. He had been in charge of a munitions factory in India and that was, as far as I was aware, his only qualification for the job. But he certainly brought with him his instincts to command.

The other representative from the NRDC was a partner in a City firm of accountants, who always simply consented with the brigadier and made no pretence at understanding anything remotely technical.

However, the accountant was keen to stress, (along with Holman and the other directors,) as a first priority the need to reorganise Technograph's accounting arrangements, which had until then been provided by Henderson & Spalding's staff. The accounts were indeed in a mess but the job of getting them in order grew in strict accordance with Parkinson's Law. One, perhaps inevitable, result was the cutting of budgets throughout the firm. My requests for even minor investments in additional production equipment and for a development fund were constantly put off until such time as we could pay for it by profits, after we had recovered the losses accumulated in the books. Expenditure on development, they said, had to stop.[3] We were already so far advanced over the whole radio industry that we should better try to make money by what we had already perfected, particularly the printed circuit boards which industry apparently now wanted. That was the NRDC line.

This opened an unbridgeable gap between the whole Board and myself. No compromise could be found and the quarrel became deep and bitter. The Board interfered in my management of the factory and dismissed the administrator and the two highly qualified engineers who were my first assistants. They were the only Jews in the company other than myself: both were refugees from Vienna.

I took up the flight on all levels. I openly defied the Board on the dismissals, alleging breach of my contract as Technical Manager and threatening to publicise what I considered was scandalous behaviour. I succeeded to the extent that the salaries of my assistants were paid until they had found jobs with other firms; this was not difficult for them. Other companies eagerly hired them as a means of obtaining technical know-how for which they would otherwise have to pay Technograph much, much more than what it cost them to employ my ex-assistants.[4] A pyrrhic victory for the Board.

The basic principle of concentrating on production for the growing demand of the market, namely for printed circuit boards, was not actually opposed by me, even though the NRDC refused to provide any money for expansion of our production facilities; I felt only that to carry out this task of routine production did not require my full-time personal attention in the factory.

Stopping development however was a much bigger issue. In order to understand the case, as I saw it, I have to go back more than a year to the period in which the negotiations with the NRDC had begun in earnest. I had already been in America with our exclusive US. licensees, had studied the American scene particularly from the viewpoint of the licensing policy of the American company and the reactions and attitudes of the American industry and armed forces. I had also travelled several times to France and Germany. A considerable number of my more than two dozen British patents on various printed circuit subjects had already been published.[5] I had even begun to write articles and had become known as an inventor and pioneer. The most important event to put the Foil Technique and me personally on the map was my paper "Printed Circuits: The General Principles and Applications of the Foil Technique" which I had read at an overcrowded meeting of several hundred engineers at the British Institution of Radio Engineers, and for which I was later awarded the Marconi premium of the Institution for the best engineering paper of the year.[6] The stir caused by this lecture in engineering circles within the radio industry, our coveted customers, a stir, moreover, associated with my name, ran very much counter to the policy of the Board of Technograph and the NRDC. They had always kept my name out of any publicity. All rights and all merit had to reflect on the company: individual boffins were to stay anonymous and in the background.

My emergence from this position gave me my main psychological support in the fight which I waged alone in the Boardroom against the NRDC's decision to stop all development expenditure including time spent on such jobs. No allowance was made even for inventions which, by the start of 1954, had not been fully developed, but on which the patents had been granted and which the company and the NRDC had undertaken to develop when I had had to assign all the patents to them. My frustration at the NRDC's actions was not a good enough argument with which to fight the rest of the Board. What I tried instead was to warn them of the likely financial consequences of their direction on licensing policy, particularly abroad.

Of Technograph's directors, I alone, together with the patent attorneys O'Dell and Warr-Langton of Messrs. Sefton Jones, O'Dell & Stevens, had gone through the whole prosecution of all my patents in Britain and abroad. We knew nearly all citations of prior art and had been very careful in rendering the patents as perfect as such documents can be. I was therefore convinced that my patents were very strong indeed; so strong and novel that I had no doubt that any fair court would confirm their validity against any proven infringements. The efficiency of that legal protection depends not only on the strength of any patent but also on the power behind it, on who best can afford the enormous cost, the delay, and duration of litigation, and on who can hire the better lawyers. I felt, with Edison, who preferred to compete with infringers by proving to be better and cheaper rather than by taking them to court.[7]

My line of argument with the Board was that I could not envisage that we, a small company, could be successful in licensing my inventions just on patent rights, however strong they were. We had to continue with the development of the inventions and offer our prospective licensees the benefit of our knowhow and continuous improvements. Furthermore, we had to extend the area of our industrial interest by developing more applications. Successful licensing promised a much greater income even in Britain than our own production; The latter was necessary as well, of course, but mainly in order to acquire and prove knowhow. From abroad licensing was the only revenue foreseeable but that could turn out to be enormous. We had to retain our status as the leading printed circuit company; this could best be done, I believed, by continuing and increasing development work, and not by putting it off and losing our skilled staff.

I was particularly keen to get financial sanction for the development of my invention of integration of resistors into a multilayer circuit for which I had obtained a patent in 1953.[8] We had called the attention of many designers to this patent as soon as it came out and made a few laboratory samples. As we were already producing in quantity extremely small, finely patterned strain gauges of fairly precise resistance value, some only $\frac{1}{4}''$ long, the miniaturisation of these integrated circuits was felt to pose no further basic problems.[9] I was convinced that, with proper development work to normalise the fine two-colour printing and differential etching, with the standardisation of the multilayer materials and their testing, and further with the printed three-electrode capacitors, we would keep a monopoly position in these integrated multilayer circuits for some time to come. However, the only immediate result we got was praise, no orders; and this did not help me to move the NRDC towards agreeing to a development budget.

While I held this invention, protected by patent, as a striking example of the fundamental principle of the foil technique, I again had to face the fact that bridging the gap between pioneering a technological breakthrough and its acceptance by industry and bureaucracy takes an awfully long time. The NRDC certainly stuck to its guns: No money for development work. I failed to win over the NRDC and hence the Board, obedient to its masters. I had already assigned all my Printed Circuit patents to the company, which in turn had to entrust them to the NRDC. That and my agreement to a service contract as technical manager to last at least until 1956 had been cardinal conditions for the NRDC's loan and for its taking on responsibility for the company. By the time I had had to sign over my rights irrevocably Strong was already in such dire need of the NRDC's money that I could not let him down. At that time also I was not conscious of how inefficient and inexperienced the NRDC was and would continue to be.[10]

Furthermore, who would have imagined—I still ask myself—that the NRDC, having abandoned development on the patents and keeping only legal titles alive, would possess itself of the idea that as an official governmental body for the service of British industry it could not threaten to sue British companies who were openly infringing the patents? The political consequences of such actions for the Corporation, and the adverse commentary which they anticipated in the press for "obstructing modern progressive processes in British industry" they imagined would be unendurable.

The NRDC opted instead to rely on moral sense within the radio industry and its respect for the Corporation's fairness! This paranoia of a Government bureaucracy revealed itself only a long time later, after the battle with the Board had ended with my defeat—by which time there was little joy to be had in knowing myself to have been right.

At the end of my fight with the Board I was not released from my

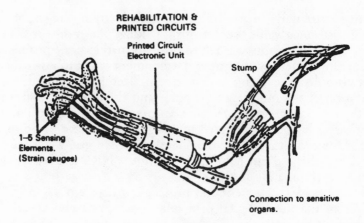

Electrolimb.

service contract, maybe to keep me silent, and so, still a director, I occupied myself with my own projects on "Electrolimb" and "Microcircuits", which were only in dreamland at that time.[11] I also worked at completing the prosecution of the pending patent applications and giving personal technical assistance to our licensees in the United States and our agents in France however much I disagreed with their actual policy. And, while the actual takeover of the organisation, which I had founded with Strong, proceeded, and while my fight in the Board went on and my departure from the Technograph scene became only a question of time, my invention had already begun its worldwide success.

The treatment which my inventions received from the NRDC has made me wonder again and again why this organisation has been so grossly inefficient in doing the job for which Parliament had created it. That job, to put it simply, should be to secure the basic development and assist in the industrial development of British inventions for the progress of British industry and the benefit of the nation.[12] In the course of my experience with its intricate apparatus I usually found its officers immersed in searching for some paths through the maze of its own rules, regulations and traditions of bureaucratic behaviour which comprised the most drastic illustration of Parkinson's Laws.

In the little time and with the remnant of energy left from the political games which occupied so much of their minds, they appeared to be incapable of thinking of promising ways of dealing with the many inventions which were put in their care.

The great number of University-scientists and engineers employed by the State and all the other public employees who produce annually a very large number of inventions are sources of great potential benefit to the nation. Their inventions fall gratuitously and automatically into the lap of the NRDC. The law provides for that where it says that all inventions of an employee relating in any way to the field of his employment belong to the employer. This in spite of some subsequent refinement is still the law in Great Britain and most so-called civilised countries. The fields covered by the inventions of the employees of the State are innumerable. Their proper evaluation and the steps necessary to produce their development are, of course, quite beyond the capability of the NRDC. Occasionally one of the patents paraded in their meagre publicity handouts may have the luck of the draw but these exceptions rather tend to strengthen the impression of the NRDC as a cumbersome angel of death for the rest of British inventions.

Most inventions are only the first steps of innovations influenced by many factors. Managing a successful interplay of these factors is the job of a champion who is immersed in the atmosphere of the innovation and whose job and money is at risk. His fate should be linked with the enterprise and not with civil service rules of promotion, impartiality, and "safety first". The inventor who in spite of having to assign his rights to the State has a natural interest in the development of his brainchild should be credited with some knowledge and creative ability in the field of his own invention; but he is usually kept out!

This principle of not letting the inventor participate in the development was much in line with the NRDC's policy towards me.[13] It also possibly had some effect on the attitude that the corporation had towards my being in charge of the works.

I felt their relief at my eventual resignation from the Board of Technograph, which happened more or less as soon as the initial contractual agreement applying to me came to an end. I had known for some time that my position with the company was untenable, and as soon as I could go without embarrassing Strong I did so.

From me the resignation signified the final decisive break with the organisation I had founded together with Harold Vezey Strong for the development and pioneering of printed circuits. At first, in 1941, we had run it as a department of his firm and we had led it for fourteen years through all the tribulations of war and resistance on very little money to the phenomenal, worldwide adoption of its technology.

I felt elated to be free again, liberated from the duty to serve the people I did not respect. They knew I would not drag them through the courts for their breach of agreement regarding my undeveloped patents. Better, I felt, not to have anything to do with them anymore and from then on I observed, as an outsider, the further fate of the company.

The Board of Technograph soon changed almost completely. Within a short span of time both brothers Strong, Major Holman, and two others died. Guided by the NRDC a reconstruction of the share capital of Technograph was carried through which resulted as far as my share holding was concerned in watering down the proportion of my holding from about $16\frac{1}{2}$per cent to $2\frac{3}{4}$per cent. The American Technograph Company started a law suit against a prominent infringer (Bendix) which took many years and was widened into approximately forty suits in various states of the United States. It developed into one of the most extensive patent cases in American legal history and cost more than a million dollars in lawyers fees. The original verdict of the court in one state, North Carolina, was so openly a misjudgment that it was corrected on appeal. Nevertheless even then the patent in question was held invalid for obviousness in one state but not in another. Referring the case to the Supreme Court was refused and the funds of the American Technograph became exhausted!

Analysis of the reason for the American legal monstrosity is a game free for all. Some just thought that the main cause was ill luck on the part of U.S. Technograph. Some will say it shows up the unfair and prohibitively expensive American legal system which gives such advantage to the large and wealthy organisation and renders it too dangerous and expensive for the small and relatively poor. Some may count the battalions on each side and emphasize the foolishness of taking on a fight with an adversary many thousand times stronger in all resources including money. I had warned, from the very start of the enterprise, not to build a licence policy on patents only and warned about discontinuing research and development, but no-one seemed prepared to listen.

In Britain the open infringement of my basic 1943 patent by industry developed partly because of the NRDC's principle of not suing a British firm and partly in the wake of the American experience. However, after Lord Halsbury had vacated his position the view grew within the NRDC that the principle of not bringing a British firm before a court of law may not be so holy. Confidence in British justice was in any case greater than in its American equivalent. Also, as time passed, appreciation of the importance of my invention grew far beyond electronic industrial circles.

The pressure on Technograph to bring an infringement action before the High Courts thus became ever stronger, with the growing age of the original patent and the success of the invention. They had to let the court decide whether my patent was valid! It took a very long time but eventually all the Courts, the High Court, the Court of

Appeal and finally the House of Lords dealt with the case. All decided unanimously that the patent was valid, that the invention was not obvious, and gave to the invention and to me personally the highest credit.

In the Court of Appeal in 1969 Lord Harman said:

There had been many paper attempts to do something of the sort before and none of them had ever proved useful or got on the workshop floor until Dr Eisler hit on his idea.

Lord Widgery commented:

It was not until Dr Eisler hit upon the idea of going outside the radio industry and looking for an answer in the printing art that the revolutionary use of printed circuits became commercially possible.

Lord Sachs explained:

Before 1943 no-one in the electronic industry used the relevant method for preparing a circuit board network: the idea of applying the relevant method to making circuit board networks originated with Dr Eisler. It was Dr Eisler who through his specification introduced the application of this method to the industry: from this introduction stemmed the adoption by the industry of the almost universal use of the relevant method and the creation of the new industry of manufacturing circuit boards by this method: and the introduction of this method to the industry resulted in what can fairly be called a revolution.

In the House of Lords in 1971, Lord Diplock said:

The commercial success of printed circuits made by the method claimed in the patent has been enormous. Their introduction on a commercial scale in 1953 has resulted in a revolutionary change in the technique of assembling radio and television sets. It has replaced other methods of making electrical connections in about 85 percent of the total production of this great industry.

So it was that, despite the way I was treated by Technograph and especially, the NRDC, printed circuitry enhanced my personal reputation and proved to be one of the most significant developments of the century. It has even been recognised as such in some quarters in Britain. Late in 1957 Lord Hailsham made a speech at the Royal Society in London in which he not only named printed circuits among the six most important inventions of the times, he also discussed the role of inventors in a way which rang very true to me. It seemed to sum up my feelings—after 15 years with Technograph—to such an extent that I shall quote at length from the report of the speech in *The Sunday Times* of 1 December 1957.

Viscount Hailsham, Lord President of the Council, said last night: "I am an egg head by conviction, and I hope by practice. A country neglects its eggheads at its peril. . . . Let us be proud of our eggheadness."

"Both in Britain and America," said Lord Hailsham. "It is the popular fashion to ridicule and condemn what the Americans call the egghead, the man who from boyhood up deliberately and consciously pursues intellectual pursuits whether humanistic or scientific—or, better still, both—with wholehearted zest.

"An unbalanced fellow, they would say, a dreary unpractical otherworldly sort of chap, unfit for the society of successful politicians, hardheaded men of business simple soldiers, silent sailors or working journalists and all the other kind of folk who make such a botch of this practical world by refusing to think about the theoretical problems involved in the art of living. "It is time we got together. Eggheads of the world, unite; we have nothing to lose but our brains.

"It is the egghead who invents the Sputnik, who discovers penicillin, who splits the atom, who thinks of the printed circuit, the electronic brain, the guided missile in the world of science."

7: Freelancing

I had no job waiting for me when I resigned from Technograph. It was 1957, and once again I had to decide upon the field in which I could seek work. It could not be in printed circuits. It was not the service contract I had with Technograph which was the main obstacle. Rather, it was my conviction that I had pioneered that technology to such a state and to such a degree of acceptance by the industry that there was no longer any doubt that other workers could and would take over the continuation of its development. But what was I to do? I had become so immersed in foil technology, I had explored so many of its facets and had developed my engineering approach to problems on the lines characteristic of foil to such an extent that foil technology had become the only sphere in which I could move confidently.

Between 1957 and 1959 I worked freelance. I consulted small firms such as a London wire and cable firm, Hartley-Baird, as well as large outfits such as Elliot Brothers, Firth-Cleveland and Tube Investment. With the Lessings I continued the work on a special type of wallpaper heating which I had started sometime before with the agreement of Technograph.

I was involved in varied and interesting projects, but as time passed it became clear to me that much of what I was doing was not getting very far. The small firms I worked for all sooner or later gave up their new projects because, as money ran out, expenditure for anything new was the first to be cut.[1]

On the other hand with the large organisations any research which did not have a direct bearing on the immediate competitive concerns of the firm had an uncertain future. Areas of concern were defined by groups of large corporations jealously guarding their vested interests. Things outside them had an uncanny habit of ending up on the corporate scrapheap. The waste and the size of those organisations appeared to me to be in more than linear proportion.[2] To witness the stupendous growth of waste from the inside was an incredibly terrifying experience. The feeling I got that the efforts which I made and for which I received my fees were destined to eventually prove a waste of time led me to think up ideas which I felt had at least spiritual value.

If I was obliged to earn my living knowing that the work I was doing was destined not to be taken seriously, at least I could devote my spare time to activities I found more rewarding and constructive. Indeed I found enough time in these three years not only to make three inventions which I still value as among my highest achievements (although I could devote my efforts to the development of only one of them) but also to write my book on printed circuits.[3]

Of the inventions I worked on the first one I wish to describe here had actually taken root in my mind as early as the spring of 1942. While London was shaken by the Blitz I drafted an exposé on the concept and applications of my idea on printed electrical circuits. I had intended to take this to my patent agent and close friend H. K. Warr-Langton, but I missed him. He had already been mobilised for war work. Instead I met at his office old A. E. O'Dell, his partner, who was carrying on with the patent agency business. He was one of the very few British patent agents who were also qualified as American patent attorneys.

O'Dell was much impressed with my idea. After a first glance at my voluminous notes, sketches, and samples his enthusiasm broke through the formalities of business attitudes. He even teased me by asking why I had not extended the scope of my idea to small mechanical structures. Why not print aeroplanes eventually? I replied in equally jocular manner to that science-fiction question but through all the years O'Dell's remark about printing mechanical structures probably slept in my unconscious as a mirror image of the printed electrical circuit idea. Instead of the pattern of pathways of the electric current I took the lines of mechanical forces forming a pattern which might be projected into a number of two-dimensional layers. These layers would eventually become a three-dimensional structure after having been printed and produced as two dimensional parts.

When I thought about Printed Mechanics the ways in which civil engineers design bridges was the model which inspired me. The modern designer of a bridge does not use one colossal beam of solid material to span the gap, but designs a framework of, say, steel trusses. He makes a stress diagram or stress analysis of what he feels to be the most suitable pattern of latticed girders for coping with the forces to be expected. Having found the stresses in each strip and joint of the lattice he determines their dimensions accordingly so that they can take the maximum forces which may fall on them. Usually the strips are steel struts of suitable cross-sections and there is only air, no structural material, between the members of the lattice. The lattices are the only patterns of the lines of force on the girder. When we have several girders in a bridge the members of the lattices form a spatial configuration. The printed mechanic designer must see this configuration of stress lattices as a spatial configuration of lines of force. It is his task to project this configuration onto plane layers for printing and etching or for otherwise producing the pattern of lines of force as a pattern of wide or narrow foil strips. He thus analyses and composes the stresses encountered at some small machine elements layer-by-layer just as the civil engineer models huge steel frame works on the drawing board first.

Rules and miniaturised systems for simple mechanical elementary parts could be developed. As examples of such parts I thought of beams and bars used for instance as levers in engines, of flanges, stanchions and struts, webs and plates. I imagined a few standard loading conditions of such parts for which typical model lattices could be designed and computerisable design directions could be obtained.

But how would it be possible to ascertain and compare the loading stress? For that purpose I thought of photo-elasticity as a means of exploring the stress distribution in some engineering structures. I also considered the "scattered light" method of observation, by which scientists ascertained the directions of the three principal axes and the principal stress differences at any point of the model.[4] I enquired whether instead of working out the stresses in two-dimensional slices a second model consisting of cast films in the principal stress directions might be a practical approach for obtaining more information of use in the basic design of our lattices. I could not obtain a positive reply; only development work for which funds would be required could provide an answer.

As things turned out, there is no doubt that photo-elasticity in its present state of development could give only initial design guidance to the printed mechanics designer and there is little hope that a photographic copy of the stress lines of the model slices can be used directly as a blue print of a lattice. Nevertheless it can be hoped that the experience gained in the design and use of the simple basic parts given above as examples of first exercises namely bars, plates, and so on, might direct the development of the ways in which photo-elasticity could be of increasing use to the designer.

After some thought I decided to approach the large corporation for which I had done a study on high melting metals and alloys. I wrote out a realistic outline of my invention in the form of a confidential paper and sent it to the Head of the corporation's technological centre. There was indeed a fairly immediate reaction! Their patent agent telephoned me for an appointment as soon as possible; he came to London to see me and asked me to assign my rights on this invention to them against the sum of £500 and a service contract.

With the recent experience of how the NRDC had dealt with my printed circuit invention I had not a minute's hesitation in showing the patent agent out of my study. This, of course, meant goodbye to the corporation as well. Printed mechanics was obviously too hot a potato for me; I had to let it cool; perhaps somebody in a next generation will take it up and develop it to a practical stage from its present level of a science fiction dream.

The second invention I made during this period was a new, very advantageous constructional principle for a highly practical product. It arose out of an approach made to me by the well known firm Firth-Cleveland via Bowler, the chairman's secretary. He told me that the lead works of the group needed a new line. They, like all other lead works in the country were allocated a carefully regulated quota for all their production at this time. So the group were seeking a new line and wondered if I could offer them an invention leading to an increased output of lead. They would repay me any patent agent's bills and act as agents to find another interested party if we were unable to come to an agreement over anything I might invent. This seemed to me to be fair.

I proposed to develop a new type of storage battery and, more specifically, a new type of lead-acid battery which used the same chemical processes and active substances as the conventional battery, but differed from it fundamentally in structure, working, and manufacture. The invention on which the proposal was based was applicable to all secondary storage batteries, but I proposed to start with the development of the lead-acid type because this type promised greater financial rewards. Lead is also cheap and available in very large quantity.

However, despite the efforts made by me and by a number of technically highly qualified people who liked the idea of the foil battery, it failed to win over the battery manufacturers. I eventually stopped my efforts to find a way of developing my invention and I wrote it up for record purposes in the paper entitled, "The Foil Battery."[5]

If my invention had been accepted and fully developed a situation of an industrial earthquake would have arisen. The percussion of the old established extremely powerful industries based on the storage battery could be compared to the revolution which the printed circuit development brought to the electronics industry. There is no percussion imminent at present. The idea of the foil battery for the electric car rests like Sleeping Beauty waiting for Prince Charming. Will he arrive before new systems have won the car race?

Meanwhile during the last quarter of the 1950s, I made the third of the "basic" inventions conceived during the period in which I earned my living by consultancy. The invention concerned electric surface heating by means of mechanically and automatically produced foil elements and new fields for their use. I did not imagine that this invention would keep me fully engaged for over twenty years, up to the time of writing these lines. I actually intended the first patent applications for it to serve as "pot boilers."

As distinct from the foil elements produced by my printed circuit methods, the production plant I envisaged for the heating foil elements would be able to produce meander patterns from any thin foil and to laminate them on or between any insulating layer, such as paper, plastic film, or textile. No material would be needed just for processing, in the way that printing ink or etching solutions are needed for producing printed circuits, but which are not present in the final element. This new element contained only the foil and its supporting insulation. Aluminum foil was by far the cheapest readily available foil and the cheapest insulators, such as paper or plastic film, could be used with my method; furthermore, automatic machining was not only foolproof but also the least costly method for production of large quantities.

I thus believed, and still believe, that I had invented not only the technically best heating film, but also the one which could at the same time be absolutely the cheapest when large quantities are produced.

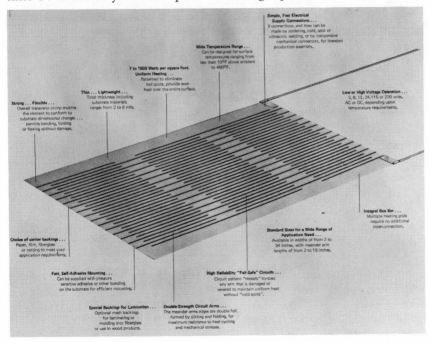

Avery-Eisler heating film.

Its cost would exceed that of the raw materials it contained by only the machining cost, which would be very small for long runs and could still be competitively priced for shorter runs. A thousand and one more applications came to my mind, promising and problematic, huge and modest in quantity and business aspects. After all, heating is, next to food, the most important need of mankind and my invention made the source of warmth, a safe and cheap dispensable commodity which could be bought like a piece of paper for packaging or to write on at the writers convenience; it could replace heat coming from an often cumbersome and expensive, high temperature source, round which men had to gather.

The principle and technical aspects of this invention appeared to me so simple and some fields of application so attractive that I first hoped to be able to sell it almost "on sight". By analogy with those of Balzac's works which he classed as "pot boilers" I hoped that the invention of the foil heating film, which appeared so readily comprehensible to the businessman's mind, could perhaps be such a pot boiler. I would then be free, with sufficient money to develop the foil battery myself. I could use as laboratory a major part of our large house which had become half empty through the deaths of my mother and my mother-in-law who had had their flats in our house. I did not talk about this plan immediately but when an occasion arose I made tentative enquiries.

The opening occurred in the offices of Mr. Bouly, an eminent patent attorney who I met in connection with my work for the Lessing brothers. Bouly knew of my printed circuit patents which I had assigned to Technograph and he had a real appreciation of the importance of my invention of printed circuits and treated me with great respect as their unchallengeable inventor.

After finishing with the subject of the patent application for the Lessings, he asked me whether I had any new inventions. I told him of one of the applications of the foil heating film—its use in the packaging of precooked food. Heating film was cheap enough to be dispensable like other packaging material and the packaging itself could be the standard used. When connected to a car accumulator or 12 Volt outlet of a transformer by means of a simple clip, an electric current would flow through the aluminium foil of the film, and create instantaneously a uniform heat of predesigned intensity over its whole area. This heat would immediately be transmitted to the food in the pack.

Bouly became quite excited and I remember the great number of prepacked foods, frozen and in cans, he referred to, which could be heated with a small transformer almost everywhere. I had already

Flexible heating film.

Stripping a foil heating film.

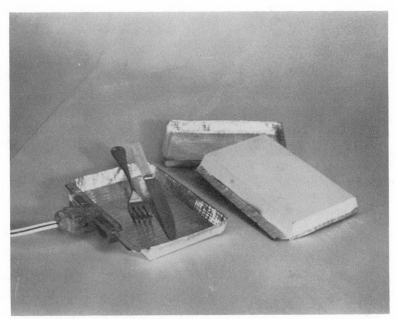

Cardboard container tray of similar construction to papier mache container.

Heating film stuck on the base of conventional aluminium container tray.

thought of some of the applications for film of this kind. It could be used for heating almost any precooked food to be consumed directly from the pack; to keep food warm, for instance in thermally insulated bags or boxes; to speed thawing of deepfrozen food; to boil or cook raw food; or to sterilize containers and foods during packing. I told Bouly that I had already lodged a provisional patent application but had done nothing further than a few laboratory tests. That did not dampen his enthusiasm, it perhaps even enhanced it.

It has been my usual experience, probably shared with most inventors, when describing and explaining their invention to somebody in a private conversation, that this somebody reacts with an appreciative kindness.

Initially I took Bouly's enthusiasm as a tribute belonging to that kind of politeness. But he showed persistent interest telling me next that he was the British patent attorney of the Minnesota Mining and Manufacturing Co., the great "3M" corporation. The Managing Director of the British 3M company would come to see him in a day or two and they, meaning the American mother firm, were looking for a new field to go into. If I agreed he would have a talk with this visitor and give him a copy of my Provisional Patent for transmission to the people whose function was to decide on new ventures. This would be a direct line by-passing various possibly delaying and interfering officials. Bouly obviously thought that I would jump at his offer. He had no notion of my esteem and expectations of a "big" corporation, which were, if anything, in nearly inverse proportion to its size. Logically and in keeping with my experience I should have politely to refuse Bouly's offer, and I had just started to thank him for the help offered and to make a polite, in-offensive excuse, when it was announced that the 3M managing director had dropped in to see Bouly. He entered the room, Bouly introduced me and started to tell him about my heating packaging. He was obviously so convinced of the opportunity he had opened for me that he mistook my appreciation of his efforts for an acceptance of his proposal. So there I was, concealing my embarrassment, accidentally stumbling into agreeing to let 3M see the provisional patent application against the promise of a quick reply.

That the reply would be speedy was all I felt I could hope for; I never shared Bouly's illusions of a positive outcome from such a submission. Moreover, while not being afraid of 3M embarking on a line leading to infringements, I felt that I had to hurry up with my patenting of the invention more fully to anticipate any possible flood of patent applications by a mass of prospective trend followers which leakage might bring about.

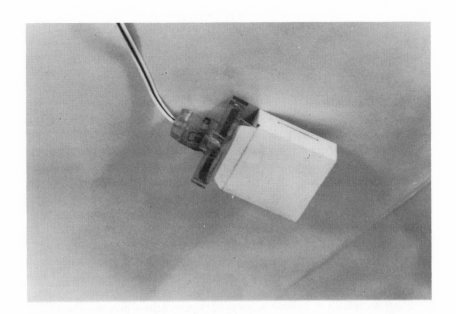

Model of fish finger pack showing the closed pack *(top)* and the opened pack *(bottom)*.

Heating film stuck on the side of a can. The views show progression from fully packaged *(left)* to open clipped-on can *(right)*.

The final reply from 3M was, not surprisingly, the usual statement that the invention was outside the scope of the organisation's interests and, in the event, it was anything but quick. By the time it arrived I had almost forgotten 3M and had moved on to other areas of research.[6]

Several years later, I did return to the selfheating foodpack, when working at Eisler Consultants.[7] We had developed the heating films for our "Hotpaks" in all the technical aspects which we could think of and could test in the laboratory. We went into market research for some prominent fields. We concentrated on two main categories: the domestic user's market and the communal user's market.

In the former, the selfheating foodpack would, for general household use, be competing with conventional means of heating food—the gas or electric cooker and microwave oven. Nevertheless, we felt that the convenience of our invention would endear it to the housewife.[8]

The communal users' market appeared, however, more promising for a less delayed success. For example, the foodpack would be useful for the motorist wanting a picnic or hot travel food, similarly it could be a boon to campers. It could also be useful in city offices or workshops which have no canteen or microwave. It could also be used in conjunction with vending machines at railway stations, air terminals, sports meetings, or launderettes. The feeding of large numbers of people is particularly suited to the concept whether it be the armed forces in the field, passengers on an aircraft, or as meals on wheels for the sick.

But to return to my rejection by 3M and the late 1950s. I was by then deeply involved in the work of designing, experimenting, and planning the basic features of the patent structure of another application of foil heating film—electric surface heating.

British provisional patent applictions I could draft myself and the fee of £1 was no problem. But within a year complete, British patent specifications and applications in foreign countries had to be lodged and this job had to be done by chartered patent agents. Even when limiting most severely the number of patents, and staggering the timing of the applications, and of the expenditure over several years I could not see any rational chance of raising the adequate funds. Even when restricting patent application to only those foreign countries in which at least some patent cover was essential for obtaining development finance in this country, the investment required for the fees of the chartered patent agents was quite beyond my means.[9]

I discussed the situation with the late Mr Warr-Langton, the then head of the firm of my patent agents Sefton, Jones, O'Dell and Stephens. They had been my patent attorneys for all the patents I made since I arrived in England in 1936. I had, of course, also brought them the account of Technograph covering almost all my British, American, and other foreign printed circuit patents, well over a hundred in total. Warr-Langton could, therefore, perhaps better than anyone else, accept the genuineness of my creations and the struggles I had been through. He was highly competent technically and saw the relevance of my new invention, some of its likely ramifications, and the immense field of possible applications. He knew, moreover, that I had no money.

What then happened was like a miracle. He made the offer that his firm would launch all the patent applications and prosecute them; furthermore I would be obliged to settle the account only after I had started to be successful with at least some of the patents. Nobody previously had given me a break like that! I had expected confidence in the invention but the credit he offered expressed also confidence in my ability to do business at least sufficiently well to be able to settle his account.

I do not know whether under different circumstances I might have abandoned an invention for which I had to procure finance from commercial sources but Warr-Langton's confidence in me prevented such an irresponsible act. The way ahead seemed clear to me: I had to bring the invention into such a shape that I could demonstrate it to potentially interested commercial organisations and then try to reach agreements with them which would firstly produce sufficient funds for settling the account with Warr-Langton's firm and provide funds for the development of the invention. My other criterion was that no agreement would require assignments of patents; they had to leave me independent and in possession of my patents.

What originally looked to be a pot boiler then turned into a very heavy slow cooker, a crockpot forcing me to submit ideas to large organisations. For me that was a rather objectionable and undesired situation, but I could not see any other way to raise money. I hoped that after some initially successful deals I might be able to play the game differently.

Designing a patent structure—as restricted as rationally possible in order to save ultimate costs—was a task I was able to do myself quite readily after Warr-Langton's offer had relieved me of the worry of repayment due at a fixed date.[10]

I selected two areas of application as those most urgently needing patent cover. The first of these was to provide cover for those features of what I called "Selfheating food packs" or the "Food Project" necessary to show to a prospective licensee, and a cover for those features which may have been suggested by possible leaks from the provisional patent sent to 3M. The second area of concern was the need for patent cover for an industrial application. This constituted an important and obvious improvement of an article already well known and used by industry, an article furthermore which could be produced in quantity with available, in expensive machines. The article was a Foil Heating Tape which would compete with heating tapes using resistance wires. The latter had been on the market for at least ten years, the market leaders being Electrothermal Engineering and Isotape.

Once I had lodged patent applications for these two areas of application I could take the first steps to seek finance. I held the view that the Food Project requried a huge investment and I had therefore to interest a large firm in it, but that the Foil Heating Tape was so simple that, if backed by only £4000, I could organise production and sale within a few months myself. At this point I needed to take stock of the situation.

Luckily I did not have to take decisions on my own. My wife was a constant source of advice and support and, through her, I had met a friend on whose judgement I knew I could rely implicitly. During the 1950s my wife had moved her research work from the Maudesley Hospital in Camberwell, to University College, London, where she had risen to Reader and then to Professor of Psycholinguistics, and to international recognition.

In the College she met Professor E. H. Thompson and introduced him to me. Tommy as he was called by all his friends, had started life as a professional Soldier; he was educated for and by the Royal Engineers and graduated at Cambridge. He then joined the Royal Ordinance Corps. During the war he went via fighting in Greece and Crete to Africa where he made the maps for General Montgomery from Alamein onwards. Some years after the war he left the army as Lieutenant Colonel and became the Professor of Photogrammetry in the Engineering Faculty of University College.

It must be a rare luck that anybody over 50 years old meets another person with whom he strikes at once as close a tie as if he were his brother, a tie which lasts the lives of both in undiminished strength and elation. Tommy was a mathematician and precision engineer of the highest accomplishment, an independent thinker not swayed by fashionable trends or prejudice. He invented and developed the Thompson-Watt plotter, an ingenious device used to draw accurate maps from aerial photographs and he pioneered the construction of precise three-dimensional drawings from photo's of historic buildings or other works of art with all measurements for their reconstitution in case of destruction or damage. His advice was of great help to us in many ways.

After a few years of consulting informally, my wife and I invited Tommy to join us in setting up our own company, Eisler (Consultants) Limited, which would be based at our home in North West London where I already had a small laboratory. He agreed and this marked the beginning of a long association; the three of us went through many ups and downs as the fortunes of Eisler (Consultants) Limited unfolded, but throughout we were all greatly helped by knowing we had the support of one another.

I was consequently with Tommy when the next stage of my attempts to exploit my patents started. After my first failure to raise the necessary £4,000, I had no further need to seek financial backing for an enterprise of my own in foil heating tapes because a meeting at the highly respected firm of solicitors, Allen & Overy, promised rapid progress with a party interested in the more ambitious Food Project. Tommy and I met a Mr D. V. Jennings, a partner of Allen & Overy, Solicitors whose reaction to the selfheating food packs was as enthusiastic as that shown earlier by Bouly. When I was introduced to Jennings only the primitive outline of the project was in my head as a formalized proposition. Nor had I already fully conceptualised the potential of the foil heating film, or the intricacy of human reaction to such seemingly advantageous yet simple reorientation of their attitudes to heat and heating.

To explain my ideas to Jennings then, I gave the example of aircraft meals. This resulted in his wishing, understandably, for some reaction from potential buyers, so he arranged for us to meet the head of catering of the British Overseas Aircraft Corporation (BOAC). I took with me a single-portion frozen food pack, a laboratory made heating film, and leads to a car battery and demonstrated the principle upon which the project was based. The reaction of the BOAC man was most encouraging and he only regretted the delay he had to expect before he could introduce the scheme.

We were, of course, delighted and Jennings started the ball rolling by putting me in touch with Deritend, a Midlands engineering concern and client of Allen & Overy, wishing to diversify and expand. The Deritend Group of companies responded most positively to the idea of the Food Project. After several meetings in London, Wolverhampton and at other Deritend factories, correspondence, tests, reports, investigations and discussions with the chairman, the financial director, and the technical and managing director of the electrical branch of the Deritend Group—the Midland Electric Installation Company—an agreement was signed. This gave Deritend an option for an exclusive licence against finance of the laboratory work for the completion of a preproduction machine and the associated product development. In addition a very satisfactory collaboration was proposed with the technical director, R. A. Joseph.

I started to believe that the long-awaited breakthrough had arrived; unfortunately this happy state did not last too long. Deritend's accountant, Mr. Civval, saw in the Food Project a possible gold mine and wished to seek the backing of a more powerful company than Deritend in its exploitation. To see my project treated essentially as the dowry of Deritend enabling her to marry into a powerful, bureaucracy was not my aim; neither was I persuaded or encouraged when it was revealed to me that Deritend was negotiating with the Bristol based Robinson Group, a subsidiary of the American Multinational Reynold Metals, the aluminium giant.

Discussions were first conducted in general terms avoiding anything remotely controversial. Then, at the final decision-taking meeting, over which the head of Robinson presided in style, surrounded by lawyers and accountants, I found myself having to refuse to agree to the terms proposed. I am sure the top industrialists present were outraged but, whatever they felt, they had insisted on absolute control without absolute obligations and I had no faith that, in such a position, they would behave any differently from Technograph and the NRDC.

So while my dealing with Deritend proved to be an important and significant episode enabling me to continue with the technical development work on my invention and freeing me for a short while from the worry of earning my daily living expenses, in the event I believe it was a major factor in holding back the commercial take off of the Food Project. This has still not reached the public it was destined to serve.

Despite the frustration this caused, there was one highly successful by-product of the Deritend episode. From my contact with Joseph of Midland Electric Installation came the possibility to salvage something before Deritend and I parted company: we concluded exclusive licence agreements on the foil heating tape project. This started within the framework of MEI in Wolverhampton and the first production was carried out using laboratory equipment we had constructed in our laboratory in London. Very soon the expectations which Joseph and I held looked to be realistic and justified to us the establishment of a separate firm. Hotfoil Limited was registered, a site in Wimbourne near Wolverhampton was acquired, and a new factory was built.

During the twenty years since the conception of the idea the Hotfoil venture proved successful and profitable even during years of general recession in the British economy. It has grown into one of Britain's renowned organisations of specialised industrial heating engineering supplying foil heating tapes and glassfibre Reinforced Polyester (GRP) electric surface heating panels to industrial plants all over the world. Hotfoil's first catalogue illustrating the comparison of round wire and foil conductors of the same cross-sectional area and weight which shows a more than six times greater surface area of Hotfoil elements concisely put the case for the foil heating tapes, and the firm has not looked back.

In the event, however, the story of the Food Project was not over: we were contacted by a number of other firms and although the Project was not in the end taken on seriously by any of them, our involvement with yet more outside agents helped me to come to a better understanding of, if not sympathy with, the ways in which large organisations work.

Among the firms who showed interest were the food and catering company Lyons, the Metal Box Company, the catering organisation Gardner-Merchant of the Trust House Group, as well as the army and the airline companies which later came together to form British Airways. All these parties outdid one another in their admiration of the ideas and the technical features of the heatable packages shown. Each assured me of the undoubted acceptance of the packages by the great majority of several classes of prospective buyers, that is, of that proportion of the country's population whose reactions they felt confident to predict. Their favourable reactions were the same as those we had experienced when we had carried out a poll among the many friends and visitors to the laboratory to whom we had demonstrated our models. But it never got as far as a market test; the whole project is still in limbo, and the question which demands an answer is: why?

Admittedly, I had taken quite a risk investing practically all the financial means I had or could borrow, and so much of my own time, that of my assistants, of laboratory instrumentation and expenses. This way I had taken the project to the point where an interested party could be offered a substantial amount: a massively strong patent cover, tested models, full information and knowhow for different models and their production, a laboratory plant adequate for pre-production and—in my assistants—the nucleus of a team to start the whole project.

But I had stopped at this point. By comparison, in getting the printed circuits going in the early 1950s, I had had to go one step further by arranging for Technograph to undertake the first commercial production and sale of printed circuit boards in this country. That had involved me, in addition to the building up of the first production plant for printed circuits, in the complex questions of timing supplies, orders, deliveries, payments, and promotional activities, which belong to the normal business of a sales organisation. That experience confirmed my belief that however necessary a link commercial activities are, they are not my job. I had little difficulty in convincing myself thereafter that there are enough gifted sales people and organisations. I neither wished nor intended to join their ranks.[11]

This psychological background to my aversion to tasks I considered to lie in the realms of accountancy, public relations, and sales or marketing, together with the difficulty I would have had in raising the money required for commercialisation, go some way towards explaining why I did not do it myself, and why I risked leaving this to others. Had I been confident of doing them efficiently we would perhaps by now find heatable packages as a standard commodity; on the other hand the project may have died anyway.

But my bypassing of the roles of other parties does not shed much light on the question of why none of the big organisations actually took up the project. They were all capable and had all claimed to be desirous of marketing the new product. They had sincerely approved it in all respects which mattered to them, but when it came to the final decision they shrank from it.

Gardner-Merchant, for instance, had accepted our proposal; the

Heated insulated pipes ready for on-site welding into continuous length.

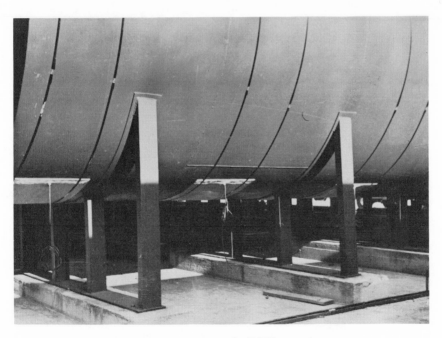

Jam tanks (1963).

Hotfoil GRP de-icing panels fitted to micro-wave aerials on the general post office towers at London and Birmingham.

Hotfoil GRP heater for frost protection of high pressure filters at a natural gas plant.

Hotfoil 'GRP' heating jackets for the prevention of condensation in storage hoppers containing carbon black.

staff and the directors were quite enthusiastic in their support. The snag for them was they had to obtain sanction from the Trust House Group, of which they were practically a subsidiary. Sadly the chairman of the Group threw the whole scheme overboard because it would involve them in activities in which they were not experienced and which would distract them from tasks already in hand.

In other organisations the project became the flag under which one group of officials tried to score points in its struggles with a rival group within the firm. The opponents stressed the anxiety, fear ignorance and resistance of wide circles of people against touching anything electrical, however low the voltage and consequently however safe. It is perhaps one of the curious aspects of such internal feuds that the arguments against new ideas from outside were more often used by the Research & Development (R&D) departments than by Sales or Public Relations groups. They suited the "Not Invented Here" (NIH) attitude of the average R&D official who seemed to resent outside inventions as they could undermine the justification of their own existence. Sales departments, on the other hand, appeared to be more conscious of the psychological obstacles they needed to consider in bringing about a sale. They were, moreover, usually in favour of expanding business with new lines.

When looking for other influences to which an outsider's invention might be exposed within the bureaucratic apparatus of a firm, we must also consider the employee's position. If the novel venture which he supports fails, he stands to take much of the blame and with it the fall of his status or even the loss of his job. If, however, the novelty is a success and he happens to have superiors the chances are that the latter will take the credit. No wonder he may be inclined not to get involved!

But, every new venture needs someone to promote it, a champion to foster and protect it. This someone needs to be constantly informed of what is happening at every stage, and to be in a position to act should the necessity arise. To have the wrong person playing this role can be fatal to the chances of the invention within the firm. Similarly if the said champion leaves a firm for any reason before a certain critical point is reached, and without securing a successor, the project will be forgotten.

Sudden changes in the financial fortunes of a company are also crucial and often unpredictable factors: the dropping of new ventures to which a firm is not yet fully committed is, of course, one of the first emergency measures.[12]

For each and all of these reasons then the Food Project remains unexploited. Nevertheless, the spin-off of Hotfoil was an early, and welcome, sign that all was not lost from the huge effort we at Eisler (Consultants) Limited had put in. While the story of the proejct was unfolding, Tommy, my wife, and I were kept busy also in other areas, trying to keep the company going by looking for other schemes to develop.

8: The Story of Space Heating

Financially Eisler (Consultants) Limited was in a worse position than it had been at the outset of the Food Project venture. From being the financier Deritend became the creditor who not only stopped paying for the expenses of the laboratory, but also requested repayment of most of the money spent. We still had money for a couple of months and we decided temporarily to drop the Food Project aiming instead at another field in which the foil heating film could be the main feature, namely space heating.

My first ideas on wall heating were included in the first printed circuit patent dating from 1943.[1] Although my actual work during the war had been devoted exclusively to creating the technological base for electronic production of the arms we so urgently needed, I was already dreaming of the comfort of wall heating as a gift which peacetime might bring.

After the war, while I was in charge of the works and of the development activities of Technograph, I did not consider any space heating project for the company. I did take on consultancy work on electric space heating panels, after I had left Technograph, but I was soon able to establish that my method of printing and etching at that time was too expensive for such large areas. I also realised that the main problem for any electric surface heating system was to prevent dangerous accidental contact between a human being and the conductors. The larger the surface areas on which such accidental contact could occur, the more costly would be mechanical or electronic protective means. I opted for what appeared to be the only reasonable proposition: to remove the danger even if such contact should occur. That meant running the whole system at a safe low voltage which would also render the system free of regulations. Such freedom would be, moreover, a non-negligible advantage when introducing a new system.

Low voltage of course presented its own problems; my system could not then be plugged straight into the 240 volts mains. What I needed was a transformer to step down the voltage from 240 to 12 or 24 volts. I chose those voltages because the public would recognise them to be safe from its knowledge of their use in cars and lorries. In

99

principle there was no problem in obtaining suitable transformers of the capacity required, but at what a price! I was shocked by my findings on this topic and began to wonder why these simple electrical machines were so expensive.

I started, as I normally do, with an analysis. Soon, it emerged that while a large quantity of transformers was produced every year in Britain, only the small ones were produced in reasonable batches. For the larger transformers the specifications of voltages, currents, and special characteristics, were so numerous and varied that there was no opportunity for industry to set up facilities of modern mass manufacture of a single type. For a batch of only a few transformers cost of production and overheads were consequently high. It seemed to me that since the components of a transformer were subject to technical design, the task of reducing their cost offered a technological rather than an economic challenge. It was, furthermore, a challenge which had some appeal for me.

If then Eisler (Consultants) Limited was to move into space heating—which was our most logical move—the transformer problem would have to be given high priority. Such work would undoubtedly be expensive so it was clear that the time had arrived, once more, to try to find outside interest. I knew this would not be easy.

It was, clear to us that only a major financial and organisational apparatus could handle any of the fields of use of the foilheating film for space heating in buildings. Beyond this our thinking identified two important points in which the basis of our negotiations would differ from those on the Food Project. Firstly, space heating in buildings by any source, including electricity, was accepted as an essential need for everyone. We were not proposing a revolutionary novelty as in the Food Project, but a new, to our minds better, and in some respects cheaper way to provide for this need. We should not therefore need to convince any potentially interested party that there would be a market; the demand was there and we hoped to offer a highly competitive and attractive way to satisfy it. In the second place, we could add an important item to the overall package compared with the Food Project. We were prepared to build a production machine ourselves and sell it at a fraction of the price any well known manufacturer of such machinery would charge.

It was undoubtedly a risky undertaking for us, but we felt we had acquired so much knowhow from the building and running of our laboratory machine that we had a good chance of success.

There remained then the question of whom to approach. I decided to take advice from Lord Sempill who had taken an interest in my work on electrolimb.[2] As well as being a peer of the realm he had an engineering education and showed great interest in new technological ideas. We had met a number of times during my years at Technograph via my then employer Harold Vezey Strong. Since then I had remained in touch with him, despite his ultimately unsuccessful efforts to obtain government help for my printed circuit work during the war (through his contact with Sir Henry Tizard).

An opportunity soon arose through a friend of Lord Sempill, who had a share holding and some influence in the business of James Williamson. They were not interested in a wall heating project, as I had hoped, but in floor heating. Messrs. Williamson had a large factory in Lancaster, where they employed about 4,000 people in their capacity as Britain's leading manufacturer of linoleum and plastic floor coverings. The firm was having to do some rethinking at the time. The rising living standard among the British population manifested itself increasingly in a preference for wall-to-wall carpets instead of linoleum. The firm therefore showed great interest in the possibility of a new type of floor heating.[3]

That floor heating could be a most agreeable form of heating had been well known since ancient times. The Romans had heated floors in their thermal baths throughout their empire. However, in later years when modern builders of office blocks used electric heating cables buried in the concrete floor to provide "night storage" floor heating, the advantage of running it on the specially reduced off-peak tariff offered by the electricity supply corporations blinded their judgement. Their night storage installations gave trouble. In the morning the floor was too hot and in the afternoons too cold; no additional heating had been provided. Office workers rebelled and floor heating earned itself a bad name. Nevertheless, in our survey of the space heating fields, floor heating ranked second, close behind wall heating. If properly controlled, we believed, it could give most excellent thermal comfort in living rooms, nurseries, schools, and offices, particularly if air movement by draught could be minimised.

I based our system on one of the main characteristics desired for space heating, namely the quick response. I proposed covering the whole floor area. The negligible mass of the foil heating film kept its inertia equally negligible. Furthermore the position of such virtually inertia-free element as near to the surface as was practically feasible allowed the occupant of any room to adjust the energy supply to the film over a sufficiently wide range to provide optimal comfort within a few minutes. If properly developed a transformer feeding the film offered easy and cheap control facilities.

The managing director of Williamson, Mr. Martineau, together with the technical director, came to our laboratory where I showed them the foil heating film and the laboratory plant. They understood the significant points immediately and did not think that a production plant based on our laboratory practice offered any further major problems. My proposals had, on the whole, a very good reception. Only the problem of cheap transformers—the solution to which I could not substantiate at that time—was held against it. As far as these devices was concerned all I had to back up my claims were ideas and basic calculations.

The advantages of our heating film over standard cables were numerous. "Normal" floor heating installations are usually feasible only for new buildings when they are put in during the construction of the building. Our floor heating could be installed in all types of building, securing for it a very much larger market. Our product was designed to run at a safe low voltage, for which transformers could be made available at half traditional costs, with no risk of electric shock or fire. Normal installations were labour-intensive making them expensive; by comparison ours were simple and inexpensive: a quick do-it-yourself job.

When it came to running costs our heating had the advantage of quick response, which was not wasteful and which could even be run at times of high tariff. Our film ensured low intensity heating, avoiding the high temperatures of single cables, and we could provide higher heating rates per square meter. With a heating film the mean surface temperature could be well below 40°C, whereas heating cables can have steep heat gradients between different sections with the associated risk of damage to floor coverings. Moreover, any local damage to our installation would leave the remaining area unharmed, unlike the situation with wires.

Our floor heating could be readily installed, one room at a time, in existing premises without upheaval. Any prospective buyer could therefore try it out in one room first, and introduce it gradually wherever and whenever he desired (and afford) it.

The heated floorcovering or underlay would not be a fixture to the building. The owner could take it with him when he moved. It would therefore be an investment not only for landlords but also for tenants.

It would be so absolutely safe, even on wet floors, in bathrooms and nurseries, that it would be free of the Institute of Electrical Engineers' regulations, would require neither special precautions nor conditions for installation, neither maintenance nor care in running and would give at the lowest cost the highest comfort of any known floor heating.

I placed these facts quite openly before the Williamson people and it was perhaps my sincerity which won the day. Whatever it was they accepted the scheme including a contract for the development of the transformer and the building for their use of a production machine, subject to our carefully following their instructions. The chief engineer, Geoffrey Alston, adopted the proposal wholeheartedly in all its aspects and became its flagbearer within the factory at Lancaster. He visited us frequently and gave us a lot of guidance for designing the production machine. Our part of the development work on the transformer was the design and trial in experimental laboratory conditions of the shapes of core and coils and of methods which would lend themselves to their automatic mass production and assembly. This part delighted Alston.

His intention was for Williamson to produce, in Lancaster, the transformers as well as the floor covering with the foil heating film. Our transformer, was being developed on lines very different from the production methods used by established transformer manufacturers, so that a new set-up would be required anyway. The resulting turnover—even if there were initially only a relatively small number of installations—would be large enough to save the Lancaster factory from the consequences of the slump in lino sales. Once started, we believed, the sky would be the limit.

We were happy and went ahead full steam. The expression "steam" is indeed more appropriate for the rear end of the production machine. We designed the paper adhesive and foil handling and pattern producing front of the plant. We organised the making of all parts and had to assemble these parts to a properly working machine for which Williamson would provide a huge steam drum for drying or gelling, and the rewind mechanism including accessories. Williamson would also arrange for all supply services and help us through the support of their large workshop, their electrical installation section, and other facilities.

Looking back, it appears almost a minor miracle how relatively smoothly it all went. After all, we were just a few engineers in a laboratory which comprised only a few rooms in a private house in London, and we were building a first, quite novel, and relatively major production plant for a Lancaster factory several hundred miles away.

Moreover the machine was completed in good time and tested in experimental and proper preproduction runs. Following that, further tests including field tests of the preproduction material were arranged. The heatable underlay material was originally tested under a great number of floorcoverings laid over a boiler house and in a passage where many stiletto-heeled women walked on it daily. Such full scale tests lasted between several months and two years. All the occupants of the houses where tests were carried out spoke highly of the standard of comfort and cleanliness and were well satisfied with the heating. The vinyl coverings, which were of delicate colour, stood up to the trials perfectly. In one house, where there were children of school age, they preferred to play and do their homework on the floor!

Unfortunately, despite the existence of the machine, the promising state of our transformer development, and the highly successful tests, the whole scheme came to a full stop. The reason was that Williamson could not stand the financial strain of its declining linoleum trade, and was forced into a merger with the great carpet firm Nairn. The resulting company, Nairn-Williamson Limited, was 77 percent owned by Nairn, with only 23 percent of the shares held by Williamson investors; all decision taking was thus controlled by the Nairn executives, who determined policy from their head quarters in Kirkcaldy, Scotland. The new firm had no use for two managing directors or two chief engineers and it was not long before these Lancaster people were laid off. With them went all their schemes: Floor Underlay Heating and Transformers were prominent causualties.[4]

So with deep regret by all who had in any way taken part in the project, our agreement was terminated, the Lancaster machine was dismantled, and as many of its parts as possible were disposed of by the new masters. It was, indeed, a sad end to a hopeful start and to the several achievements attained.[5]

Would another giant firm in the carpet trade take over the scheme? I thought it appropriate to ask Alston. After he had lost his job he joined our team for a few months and from his knowledge of the trade remodelled the proposal a little to best fit the giants he knew. He went to Ireland to see Cyril Lord, the carpet magnate, and approached David Bridge & Company in Rochdale—the machinery manufacturers who had supplied Messrs. Williamson with a large linoleum making plant. While he was thus engaged I went to Paris to see Sommer, the French carpet giant. In all cases the offer was rejected. They accepted all my claims and Alston's arguments, but were reluctant to touch anything new and different from their own lines in which business was quite good at the time.[6]

That was the end of our links with Williamson and after a short time Alston disappeared from the scene as well. Our scheme of Floor Underlay Heating, however, is not dead: it is only shelved. It needs little change. Perhaps when the electricity tariff becomes comparatively less punitive, its attraction to the public will bring it into life again.

Unfortunately, a mass use also vitally depends on a cheap mass-produced transformer being available and there was and is no established or potential enterprising transformer manufacturer in sight.

Although we had avoided any publicity about our floor heating projects, our work for Williamson had become known. It aroused interest with some engineers concerned with keeping airstrips operational in very cold climates, or with preventing traffic troubles on bridges and other sensitive stretches of roads through icing periods. On request we submitted our ideas to the relevant authorities, but little happened after initial enquiries.

As far as space heating was concerned I had, in any event, preferred the idea of wall heating. The basic concept of heated wallpaper had been with me for a very long time. The general idea is not an invention I have ever claimed, but it had always attracted me and, whenever I could, I went back to it, fascinated by the various problems and opportunities connected with it. As already mentioned I had described in my February 1943 patent application for printed circuits how the foil technique could be applied for heated wallpaper production.

Subsequently when I consulted the Lessing Brothers, after I had left Technograph, I started pilot production using my invention of patterning the, at the time, very cheap paper-backed tin foil, according to the method I had described in 1947.[7] The Lessings secured the name Space Heating Limited for a company they founded after I had supplied them with some full size prototypes that worked well. They then issued a press release.

The Heated Wallpaper was to consist of a patterned metal foil with integral connections laminated between two lining papers. Such a conductor, though thin, would carry a substantial current because its flat form would promote loss of heat by radiation and conduction. The conductor was to be insulated and protected by a covering, for instance of varnish or plastic, on which powdered metal oxide could be dusted to increase radiation; in the case of aluminium the conductor would preferably be covered with oxidation for the same reason. I envisaged that it could be fixed like any other lining paper, and anyone able to hang ordinary wallpaper would be able to deal with the heated variety with no knowledge of electricity. Like other lining papers when hung it could be coated with paint, distemper, or decorative wallpaper. The electricity supply would come at 24 volts from a step-down transformer, so that even accidental danger would be eliminated, and, as a further bonus the installation would be outside the regulations of the electricity authorities. It would be universally usable in existing and in new premises, would permit variation of heat, both regarding special distribution and timing, and would compare favourably with other electrical central heating systems when it came to cost. The initial outlay to the do-it-yourself expert could be brought down to a very low level.

Over and above this, heated wallpaper would, I was certain, provide a most efficient conditioning for the entire enclosure. As a major part of the whole wall surface is heated, this heating need only effect a temperature rise of a few degrees above the desired comfort level, so that the wall would not feel hot to the touch, usually remaining in the mid-70s (degree Fahrenheit). There need have been no hot spots, no damage to furniture, no drying of the air, no dust, no noticeable air currents, no streaks and no cleaning of parts.

Then the most exciting part of this type of heating for me was the physiological effect. In a perfectly comfortable room, there would be none of that feeling of heat and dryness associated with other types of central heating, and it seemed likely that people unable to sleep in centrally heated rooms could be quite happy with my system. The point was that instead of heating the human body, the surroundings were merely to be kept at such a temperature that the body does not have to give up all its heat to them. The only problem might be the exclusion of draughts, but that was a problem with a variety of solutions.

With cooler air there would be less drying out of furniture and woodwork, and less shock when going out into the cold. The whole idea of a room could be comfortably used with a system which would be safe and maintenance free. There would be freedom of movement and utmost saving of space. Finally, it would be possible to have a choice of instantaneous operation—switching on when needed—or of off-peak storage heating: the latter could be achieved by covering the heated wallpaper with a thin layer of foamed plastic.

A reporter from the British Sunday paper, The *Observer* came to interview me at Exeter Road and wrote a favourable article in the 22 January 1956 issue on the development.

Space Heating by Electric Wall-paper

By Our Scientific Correspondent

A British firm is developing a new method of heating rooms, using a form of wall-paper containing electrical elements.

A good deal of development work has still to be done on the new material and it will not be on the market for at least a year. It should be cheaper to install than existing forms of electric heating.

An additional advantage will be that a room will feel warm only a minute or two after the new heating material has been switched on. This is because it will produce a gentle radiant heat over a very large area.

In the Ceiling

The new material is derived from the technique of printed curcuits, in which wires are replaced by flat foil conductors bonded on to a backing of insulating material, and produced by what is essentially a printing process. Dr. Paul Eisler, who invented printed circuits in 1943, is technical consultant to Space Heating, Ltd., the company which has been set up to market the new product and is supervising the development work.

Many possible ways of using the new material are being investigated. It could be applied to a wall or incorporated in walls between rooms, mounted on a screen, produced in the form of a portable heating panel, or—and this seems to be the most probable first method—applied to the ceiling, thus warming the whole room from above.

The material will consist of a dense pattern of thin metal foil conductors "printed" on a special backing paper. It can be decorated or covered with ordinary wall-paper. When it is installed, an electrician will simply connect the material to the electricity supply, and an ordinary switch will turn it on or off.

The press release and this article caused immense excitement in the press and broadcasting companies all over the world. For a week I was beleaguered by reporters and incessant telephone calls from as far away as Japan and even Celebes, not to mention the Continent and America. The response of the public had its funny side too: three dozen householders in my neighborhood offered their premises for free trial installations, and cartoons appeared in the papers under the heading "Switch on the wall!" In addition hundreds of letters came in with offers of agencies, licensees, and from job hunters. The enthusiastic mass reaction triggered off by so little effort indicated that a heated wallpaper, if properly managed, could have enormous success.

It was clear to the Lessings that the project needed finance and organisation far beyond their own relatively modest resources. It appeared however that the interest generated by heated wallpaper had not prevented prospective financiers from getting cold feet and I could not wait forever for a Messiah from the City. Even though at the time, in 1956, I was not in the financial difficulties I was later to face I still had to earn my living. As a result our ways parted, amicably but unavoidably.[8]

The next time I had a chance to carry out major practical work on the subject of heated wallpaper came several years later through negotiations with the de la Rue group of companies. Armed with the results of our study on the heated wallpaper project I approached the Reed, Wall Paper Manufacturers, and Formica. Responses were mixed: for Reed the project would take too long to reach the minimum tonnage of paper sales which their investment policy required. Wall Paper Manufacturers expressed interest and wrote a favourable letter in the first instance, but subsequently dragged their heels. Some time later I learned that they were in some way taken over by Reed.

With Formica, the reception was quite different and positive. Some old timers remembered that I had helped them and had moved them into producing copper foil laminates for printed circuits after the war (a most profitable business, worth millions of pounds). This track record no doubt contributed to their giving my proposal of wall heating careful consideration. The outcome of this was not quite what I expected, but it was still very positive. Formica constituted one of the three companies belonging to the de la Rue group of companies, which owned 50 percent of the shares of the British Formica organisation. The other two combines were de la Rue itself, famous as the printer of banknotes, and Thomas Potterton, the leading gas boiler makers.

My proposal was accepted by the group in an amended version: the project was to be carried out not by Formica but by a subsidiary of Potterton, thereby giving Potterton an electrical heating arm in addition to its position in the market for central heating gas boilers.

The subsidiary in question was Inferation, of Camberley, Surrey, a young firm with unused factory space. We were to build a full-scale wallpaper production machine for this site, a machine which was to produce heated lining paper.

Its specification was that of a three-ply heating film consisting of two layers of standard lining paper sandwiching an aluminium foil pattern, using a special adhesive emulsified in water. The foil pattern had all repeats in parallel and was designed for a supply of 24 volts. The design of the machine which we had made for Williamson and the pilot machine in our laboratory, served as predecessors for many features of the new design. Mr. Reardon, the chief draughtsman in our laboratory actually redesigned the new machine completely, down to most details. All parts or units which were available readymade were ordered from suppliers, and other parts were either produced by a friend who had a very small engineering shop or were produced by other small engineering firms. We made sure that these parts were exactly to the specification and accuracy laid down, and made every effort to ensure they would be delivered on time.

When the drying unit was fitted into the machine it failed. This created a problem. Although the machine was eventually finished and ready to be handed over we were late in terms of our agreement.

Heated wallpaper machine under construction.

Then we were hit by a completely unpredicted external development, which had political as well as other ramifications. What was happening, however, was to have profound implications for all heating systems: natural gas had been found under the North Sea and was to be piped to Britain.[9] The blow for us was that instead of delivering the gas to electricity power stations to be converted there into electrical power, as had been hoped by the whole electrical industry, the government decided to sell it, in place of town-gas, directly to the consumer. To this end the government started to have existing town-gas boilers and other equipment converted to burn natural gas.[10] To those of us with an interest in electrical heating it had dire implications on our work for Inferation.

Simultaneously a financial crisis developed at de la Rue, for reasons which had nothing to do with our project. But it created a lobby interested in saving the role of Potterton as the leading gas boiler makers as a way of saving their own skins. If there were to be casualties, it was argued, our project should be abandoned as a first step. But however fictitious, such a step dovetailed nicely with the new gas-based fuel policy of the government. Moreover, added to this fiction was a very real financial crisis at de la Rue. We had to realise that the de la Rue group wanted to be released from their obligations under the agreement with us. The legal battle which followed lasted for several months and the mounting expenses were a far heavier burden for us than for them. So we agreed to a settlement. We got the machine back and the de la Rue group was released from its committment. It was a heavy fall for us.

It was not the loss of the de la Rue connection and finance that I most regretted; I was more worried by the changed atmosphere in the wake of the government's new fuel policy.[11] Central heating with North Sea gas was fostered, while at the same time the use of electricity was restricted by high tariffs and an effective public relations exercise. Under such circumstances I did not think that I could find another sponsor for the heated wallpaper project in Britain. But I still believed in it.

The steep fall back to square one was not only painful, it was also potentially very dangerous for the morale of our crowd, relying on our pityfully small finances. Fortunately the danger to our morale did not manifest itself and I think that was perhaps due to the fact that everybody in the laboratory knew in full detail what and how everything happened. I had kept them fully informed and despite our misfortunes they shared my confidence in the heating film.

Moreover the story had not yet run its course. A Mr. Gunnar Klein from Norway appeared on the scene. He was on the look-out for projects suitable for his, at that time, idle factories in Halden, near

Oslo. Such projects had to combine various features: the promise of profit and of employment, and the ability to make use of some of the empty premises in Halden to which he had access. He claimed that he was able to acquire plant, knowhow and all necessary accessories, provided his investigation of the technical and commercial aspects of the project were positive.

It must have been difficult to imagine a project fitting better to his needs than our foil heating film enterprise. Norway then had practically no coal, gas, or oil and was keen to expand the use of the electricity it generated in abundance from water power. Energy based on electricity, so out of favour in Britain, was naturally the cornerstone of Norwegian energy policy. We could, moreover, offer Klein a turnkey position: a plant ready and tested, unique knowhow and experience, and most reasonable terms, resulting from the urgency of our need to dispose of the machine. It still sounds like a fairy tale: a highly improbable conjunction of needs; and it went like a dream.

Klein organised a thorough investigation, to be carried out by Norwegians. Yoetun, the head of the finance house, their lawyer, and the chief engineer of the leading Norwegian converting machinery firm, all came over to inspect the machine in Camberwell, and to discuss the technical, financial and legal points.

An agreement was signed, to our great relief, just a few days before we were due to vacate the factory in Camberwell. The Norwegians went ahead with their plans to form a company, Norwayfoil S. A. The machine was transported to Halden, positioned, assembled, tested, and run in under the supervision of Rearden, whom we sent to Norway for that purpose.

Having ensured their ability to produce, Klein engaged a manager who, in turn, engaged a small staff. They rented an office in Oslo, had publicity material printed, and made their first official contacts with potential buyers of heating film. It was thereby learned that NEMKO (the Norwegian standards authority) approval had to be obtained for any application of the heating film before any sales could take place. In view of the quality and safety features of the film Klein was confident that the complex path through numerous bureaucratic steps leading to approval could be fairly quickly negotiated.[12] In the event this confidence was completely misplaced. For us in London the need for such approval came as a shock. We only then realised that equivalent approval was necessary in all other European countries, and that we would therefore have to leave trade with the Continent to licensees who could better afford to invest the time, cost, and effort to deal with respective national approval authorities, before they could expect any returns.

9: Expansion and Final Success

The situation after the end of the wallpaper project with Potterton called for another reappraisal. At Eisler Consultants we had so far carried out the functions of invention and development of various promising applications of the foil heating film, as well as of the film itself, and of the method and means of its production. We could offer everything up to the production and marketing stage of these applications. We had added to previous efforts of design and testing, the manufacture and installation of full-scale production plant for the licensee. We had not ventured into the commercial side in the way that many other enterprises might. We left it to the licensee to carry out all such aspects, including market research or market tests, publicity, and the running of a sales department. We appreciated the necessity of these links between producers and buyers but, as explained previously, we relied on others to provide them. However, the experience we had with Williamson, and Potterton indicated that we should at least carry out the production of the heating film ourselves, certainly during the early stages of a project.

This decision required a certain type of reorganisation of our small team and laboratory, changes which were perhaps in any case overdue. Fortunately this did not cause any hardship to anyone. Rearden, our head draughtsman and designer of production machines was pressed by his family to move to the West Country, and his assistant mechanics had already left. We were thus able to cut back on these activities. Rearden's departure was followed by the appointment of Dr. Vic Mazza as head of the production unit. Parallel and more or less simultaneously with these moves we shifted the pilot and laboratory machines to premises in Villies Rd a site some ten minutes away by car. This move virtually trebled the space we had available in Exeter Road, space we intended to use to increase our production facilities. We aimed at supplying foil heating film to any new enterprise hoping to take up applications of the film commercially. Such an enterprise would not thus be faced with the cost and delay of obtaining and running a machine of its own. We, on the other hand, could escape from the roles of manufacturer and supplier of a uniquely specialised and relatively complex type of machine; we might then

112

concentrate on the development of applications of the heating film and its production. As production did not fit well into the framework of Eisler Consultants we founded Foil Engineering Limited for this purpose—a legally independent corporation but in practice one which resembled a siamese twin of Eisler (Consultants) Limited.

Provisions necessary for producing the film reasonably cheaply and in viable quantities at our refurbished pilot plant were simplified and accelerated by the progress which had been made in the development of dry film adhesives. Hot melts and pressure-sensitive adhesives, suitable for most of our heating film specifications, had become available on reels, and on one or both sides of a paper or plastic film carrier. We made use of these eagerly and eliminated the wet coating and drying stages from our pilot plant. This change turned the pilot plant into a cost effective production unit readily able to run small quantities as cost efficiently as medium-large quantities. We thus graduated into proper manufacturers, but still without our own commercial means of selling.

We could not, and did not, make any great effort to get orders but we were, of course, glad to receive them. In view of the novelty of the foil heating film, and in view of the general ignorance of the trade about it we assumed that the principle source of orders would be licensees interested in launching a new project using the film. Certainly my inventions and patents of various applications provided a wide selection of such projects. Some were major projects, some minor, some turning out to be commercially successful and profitable to us, some less so. Over the years the number of applications on which we worked grew to such an extent that I can mention here only a selection.

Usually we had a number of projects proceeding simultaneously, each spread over different lengths of time, occupying the laboratory and being staggered quite irregularly. It is therefore difficult to give a strictly chronological account of our dealings with the various applications: such an historical account might confuse the description of the contents and the basic ideas behind the applications. I prefer to give brief descriptions of particular projects in order of the practical realisation or commercial acceptance which they have so far enjoyed.

I shall start then with our work on ceiling heating. Since the 1950s electrically heated radiant ceilings have been widely used in Britain, mainly in new houses. At that time the Scandinavian firm ESWA and Imperial Chemical Industries (ICI) practically shared the market. ESWA used a heating element consisting of very thin, wide leadfoil strips between plastic films, while ICI produced its Flexel film of carbon particles embedded in a silicon-elastomer. It was axiomatic to us that our foil heating film, with an aluminium foil pattern of only 0.125 inches armwidth was technically superior to both these films and might even in modest quantities, be produced more cheaply, but we did not feel confident to challenge these big organisations.[1] However, out of the blue, a Mr Crowder got in touch with us. He had been in the ceiling heating business in connection with Flexel and a Norwegian competitor of ESWA, and was enthusiastic about our element. Under the name of Alumec he managed the impossible: he competed successfully with the two market leaders. We assisted him in all but financial aspects, helped him to get the necessary certificate and did all the development work for him. It was our first fully realised and accepted space heating project. We could finally claim a commercial success.

Unfortunately our entry into the electric ceiling heating business happened at a time of increasing difficulty for all electric space heating in Britain resulting from rapidly rising electricity tariffs and the tightness of housing finance.[2] While the Alumec system was highly satisfactory and profitable for us and for Crowder while it lasted, this was not for very long. Like ESWA and ICI's Flexel—and simultaneously with these big enterprises—Alumec went down. Consequently Crowder had to cut back his activity in this area to an almost negligible level.

Because new fuel policy imposed high prices for electricity, other than under a night current tariff, there was a drastic reduction in demand for all systems for heating whole spaces electrically. This price barrier, however, disappeared and was indeed reversed in the case of localised electric heaters of, say, six or more square feet in area emitting low temperature radiant heat to provide a comfort zone of six or eight feet within their beam. To exploit this advantage of "spot heaters", we developed and designed wall panels and free-standing, moveable panels, rug underlay heaters, foot mats and strips and, with new designs, picture or poster mounting boards and heat fittings to be fixed to or hung from the ceiling. The purpose was to provide comfort for a person in the relevant zone. They had an average energy consumption of about 500 watts, and could, like light bulbs, be connected to the 240 volt mains, even where there were severe limits on power consumption.

We paid great attention to the aesthetic character of our heaters, to the materials and frames, aiming to prevent hot spots from occurring and any live parts from being exposed. None of the heaters we developed for position on the wall, on the floor, or fixed to or hung from the ceiling, required space in the room; moreover, none required installation by an electrician or other expert. They were very light indeed and any handiman could handle them competently, whichever design was chosen. All were moveable; no maintenance or servicing was necessary; the only expenditure was the initial outlay and modest electricity charges while running. They were produced and marketed under the trade names Environite and Slenderad.

We had become medium quantity producers of film instead of remaining only licensors. We were driven to the conclusion that producing the foil heating film was much more profitable than licensing the inventions and technical knowhow on which they are based. I was relieved of the constant financial worries which had been with me almost continually since I had left Technograph. Foil Engineering reached the stage where we could up-grade the pilot plant into a small but fairly efficient production unit, while experimental, model and prototype work continued at the laboratory in Exeter Road. We could now afford a development budget enabling us to carry out work when we strongly felt this was worth doing.

The successes we had with Alumec, Environite, and Slenderad were followed by others. Some where we deviated from the usual design of the heating film, others which were concluded because we could offer a strong patent, an attractive design, or even a complete novelty. In all cases the cost of the heating film was so low that the price we charged proved attractive while still permitting us to make a reasonable profit. The movement of our balance sheet from the red into the black took a very long time, but the trend was unmistakeable even though (or perhaps because) we did not try to expand, and did not ask for or use credit from banks or other financial institutions.

We remained a tiny group continuing research and development in many aspects of the foil heating film and its various applications, as well as involving ourselves to an ever increasing extent in its production. All investments we made were into improvement of our production facilities.

I was, of course, conscious of the consequences of keeping our organisation small and manageable, perhaps, some might say, too small. It meant limiting the efforts of exploiting potential markets and inhibiting growth in the use of the foil heating film. It even meant that knowledge of the film's existence and capabilities was restricted to the few people who learned of our work by word of mouth. However, we stayed a group of colleagues and personal friends and never took on the characteristics of the standard employer and employees organisation. We succeeded in remaining masters of ourselves and independent of the whim of others. We had no problems of personnel management or difficulties arising from chains of command or from bureaucratic entanglements. My policy was to stay as a small company not because I thought small was necessarily beautiful, but because I believed it to be healthy.

Our tiny organisation never had a commercial or sales department. We could not afford it. We sold either directly to accidental contacts, or through small firms launching a new product based on the foil heating film and on our product design, or through people working as self-appointed agents or intermediaries. The fact that we had no sales department certainly caused us difficulties when it came to the extension of business. Nevertheless we preferred to cope with these problems as they arose rather than to grow out of control.[3]

Our customers often came to us attracted by new technically superior products, which appeared to them to be welcome and possibly safer additions to their sometimes precarious trade.

The ups and downs of the British economy hit small businesses particularly hard. Our applications were generally not hit directly but a catastrophe in the trade of our customers was bound to have a knock-on effect on us. Nevertheless, as long as our new products were available to the public, sale of them constituted a sizeable part of our income, and the great number of products of our design which were launched successfully made the failures financially bearable. We were fortunate not to have technical failures in our design rebounding on us, although we introduced our own modifications and improvements in response to experience in the market.

Over recent years, during which time development of our organisation to an industrial enterprise had top priority for us, we worked to varying degrees on scores of products. Some of them were obvious, straightforward applications of the foil heating film with essentially only problems of technical detail; others were complete novelties, conceived by us or by people who came to us as, or on behalf of, prospective customers. The latter group of projects provided us with the greater challenge and associated interest, and some examples of them are worth exploring in further detail.

From the very beginning of the heating film project in the late 1950s we spent a great amount of effort on applications for use in the motor car. Among the first applications we had in mind were ideas for heating inside the car, and for certain car accessories. The field of car accessories is wide and very competitive. We worked on a number of products within this field, including footheaters, heatable seats to be fixed to and detached from normal car seats, methods for de-icing front windows, and demisting side mirrors, a heated cradle for batteries, and heating for the diesel fuel systems. Around each one a story of psychoeconomics unfolded which we were luckily in a position to observe quite dispassionately.

The first real hit was achieved with a latecomer, a combined demisting and defrosting rear window heater.[4] A well known firm approached us for a licence on a rear window heater; at the time there was only an unsatisfactory German type on the market. So we made a prototype and offered also to supply the necessary element. The firm snapped up our offer. It then transpired that it had handed the product to one of its subsidiaries which excelled in inefficiency. Despite this, however, the product was a great success, finding favour not only with the motorist but also with would-be competitors. One of these, a very large organisation with a smart patent department discovered a way to get around the terms of my British patent and that frightened this firm so much that they gave up the venture. The large organisation swept the market, earning millions. Nevertheless we were quite happy with the money we had earned from the rear window heater business while it lasted and we did not try to fight the battalions who had taken over the market.

We had to take a similar disinterested attitude to the various designs and prototypes of general interest requested by prospective customers. Into this category fell certain camping accessories, heatable stadium seats, tyre vulcanisers and swimming pool heaters.

As knowledge of our existence and of our expertise spread we also received some proposals without any pretence of a business or monetary motive. Some we executed, such as a heatable box powered by a portable generator for the British Antarctic Survey. It was used by the botanists accompanying the Joint Services Expedition to certain Antarctic islands during camping excursions under severe weather conditions and low temperatures. The box was used to dry out moist plant specimens collected during the expedition.

Among the strange proposals we received there were some technically interesting ones which we chose not to touch because of the organisational obstacles we anticipated from the companies with whom we would have to work. One envisaged the use of the film in the repair of water mains. To repair a pipeline the water in the pipe is frozen locally at a small distance above the break in the line to form a plug of ice. The plug is kept frozen for the duration of the repair, after being formed by spraying Freon from an aerosol onto the pipe at the required place. The evaporation of the liquid Freon creates locally a temperature of nearly $-100°C$, thus achieving rapid freezing of the water. It was proposed to use the heating film after the completion of the repair to accelerate the thawing of the ice plug and to minimise delay in restarting the flow of water. I describe this use of the heating film not because we ever dreamed of it having a chance to be adopted by such a super-bureaucracy as the Water Board, but because we were so surprised by the idea.

An even greater surprise was to come, when a consultant for a plastics manufacturer came to see us. He explained that in Britain about a million horses still needed to be reshod every four to six weeks. His firm had acquired a patent for a horseshoe pad comprising two polyurethane mouldings. One was to be stuck to the hoof of the horse, the other subsequently to be welded by the farrier to the first. This operation needed to be carried out in the field quickly and with only the use of basic apparatus such as a car battery and small transformer. If a suitable device could be made there was an enormous world market waiting to be conquered. Our visitor went on to explain the details of the mouldings outlining various intricacies which caught our interest.

The problem reminded us of another we had tackled concerning the welding of polythene gas pipes buried in the ground. Vic had had the ingenious idea of joining the two pipes by placing a special electric heating element in the interface of the pipes. For various reasons this had remained an intellectual exercise but for our customer we reconsidered it. Vic modified it slightly and we gave the information, in confidence, to the consultant. He shared our opinion that it could work. We then worked on it in some detail in the hope that we may actually reach the negotiating stage. Sadly things never did get that far as the consultant disappeared out of our lives.

Over the years at Eisler Consultants, we evolved several grand schemes for future inventions based on the heating film. Those concerning horticulture, heated clothing, and the medical field held particular interest for me.[5]

Among the orders to which we gave special attention were those involving soil heating for horticultural purposes. (Agricultural use having to wait for electrical energy becoming available). Electrical soil warming has in the last twenty years been frequently written about but has been used relatively little. For instance, in his book *Electricity in Horticulture* A. E. Canham convincingly points out the importance of the soil temperature for the very life of the plant.[6]

The first uses of our heating films in this field were for propagation and seed boxes and in private greenhouses. The competing products used heating cables to run on mains voltage. The selling price of our safe heating pads or mats was swollen by the cost of the transformer and the people who had taken up the product had not the means or courage to launch a campaign for driving home to the public the safety and other advantageous points.

When we ourselves explored the views of large commercial market gardeners running acres of greenhouses they unanimously used as only criterion for their decision their saving in fuel cost. As long as the electricity tariff was prohibitively high in comparison with the oil price we were not successful but when the price of oil rose and we had developed a particular low cost film our offer became acceptable. We witnessed the first slow beginning of one of our grand schemes, Soil Heating.

In the dreams I had of the "grand schemes to come" the great potential use of the film by the chemical industry[7] was overshadowed by the potential market and the socio-economic consequences of a general adoption of our ideas on heated clothing. This was not just an idle dream of mine. We had considerable experience in the complex subject from the work we had done for development contracts and projects on the fringes of this scheme. Our whole team had collaborated on the solution of the various speciality heated clothing problems we encountered on a number of jobs which we carried out as development work in our laboratory over several years. That occupation gave us some insight into and appreciation of the technical requirements.

We were, for instance, granted a development contract from the Admiralty to investigate the feasibility of heated clothing for deep sea divers. We concluded that we had to find first of all a thermal insulation material still effective under the pressure and helium-oxygen atmosphere in question. Dr Franz Breuer, the chemist in our team, did most of the work resulting in some positive findings and a good chance to solve the special heating problem for the deep sea divers.

Unfortunately no solution has even as yet be found for the danger to man when ascending from the high pressure of deep sea to the atmospheric pressure at the surface. This problem was, of course, not our pigeon. But it dampened our enthusiasm as we realised that however successful our finding it would only have a theoretical value. It did not prevent me having wild dreams of a hybrid between a submersible and a pressure resisting coat of armour hermetically sealing the diver within normal atmosphere, but permitting movement of legs and arms. Such dreams belong, of course to science-fiction, and perhaps they can inspire a writer of such fantasies.

A much easier and more practical problem faced us in assisting the services in investigating the different heat requirements of various parts of the human body under certain circumstances. Such investigations have a bearing not only on the design of heated clothing but also on that of sleeping bags, camping, and other equipment. It is, for instance, easy to provide heating for the crew of a tank operating in arctic conditions as long as the engine can be kept running. A problem arises, however, if the tank has to hide for a long time with engine and heating off in order to avoid becoming an easy target for infrared-guided missiles. Still, gunners and others have to operate delicate equipment at a moment's notice in the bitterest cold. For such conditions the first priority is heating the hands, then come feet and face. We could offer for this purpose a highly competitive product.

The voltage requirements for body heating for shoe-insoles, socks, mittens, vests, pads, and shields are readily met for all applications where the source of current is the battery of a vehicle—be it a tank, lorry, motor cycle, motor boat, or any other.

We developed permeable heating films for disposable, semi-disposable or permanent use. They formed separate, removable linings or interlinings to be fastened by buttoning, studding, zipping or velcro to other clothing, generally between normal underwear and wind and rain proof—a light, thermally well insulating covering.

Each separate lining was electrically connected in parallel with others by a harness or belt of tape cables which had a pair of wires with a plug to push into the lighter socket of the car or equivalent socket of the vehicle.

As long as the wearer of the heated clothing was connected to the battery of the vehicle he would be warm and happy. If he walked away for a short while only the thermal insulation of his clothing would keep him comfortable but for a long absence from the vehicle he needed another source of supply.

Primary batteries consisting of torch battery cells are unsatisfactory because of their low capacity, heavy weight, and high cost. Light, quickly chargeable secondary batteries have not been available so far. Had the Foil Battery been developed or had a new system been available the electrically heated clothing would have had no further major technical obstacle.[8]

There must, however, be a psychological barrier to its use in the home or office. It is neither economically nor technically difficult to provide small units with transformers recharging simultaneously a number of readily portable batteries and dispense them like a vending machine when inserting the spent battery. Space heating of homes and offices would become unnecessary or could, at least, be very much reduced. The saving of energy and money for everybody and the nation would be crucial.

From the very first conception of the idea of the foil heating film its application as a pillow, cover, sheet or pad, as a bandage or plaster able to conveniently transmit controlled heat directly and locally to the body for any length of time was one of my most desired aims. As a layman in medicine I thought of it originally only as a modern substitute of the hot water bottle, infinitely more versatile and useful in all the respects which came to my mind, except for its need of a low voltage source of supply. Since then studies in varying depth have been made in consultation with leading physiotherapists and specialists working on various diseases in a number of hospitals in order to find out their views on the probable effects of the proposed use of the heating film.

Among the ailments the following were considered: fibrositis, arthritis, rheumatism, neurolgia, sprains, joint-and muscle-strains, and various forms of inflammation. This is of course not an exhaustive list. The patients in question ranked from those in neonatal and pediatric to geriatric wards, from disabled amputies to post-operative patients, from sufferers of hypothermia to ambulance cases requiring heated stretchers. We learned that in addition to direct heat treatment there may be uses of the film for indirect bath therapy with poultice or mustard pads, in portable individual sauna, steam-baths and wax-bath units. We were asked for designs for incubators, operating table pads and variants of the film such as enuresis sheets. The field appeared endless.

As far as a poll can permit conclusions regarding the opinions held by those questioned, the doctors understood the particular qualities of the film, important to the proposed application. The safety and instant response to controls, the availability of all sizes, the eveness of heat creation over the whole area and, not to forget, the probably low cost—all these qualities of the film seemed to be appreciated. What struck home perhaps most impressively was the fact that the design of the film for a 12 volt supply, DC or AC, permitted the use of the heating device everywhere, in premises or in the car, over any time and if desired without interval. A small transformer, built in the plug is light enough to be carried to wherever the patient might wish to move.

There was almost unanimity over the film's positive contribution to comfort and to its giving temporary relief in almost all cases in which there was no definite medical indication against application of warmth in any form. On the other hand, but as expected, most doctors held that there was no proof available that heat treatment alone had a curative effect. If combined with other treatments the share in the cure due to heat would be difficult to ascertain. That did not mean that heating is not absolutely necessary. It implies, however, more than an academic question. The investments and expenditures permitted by the authorities for the cure of illnesses and the comfort of patients are of a different order. The former meet much less reluctance to be budgeted than those for providing "only" comfort and relief. A luxury such as comfort—of only temporary benefit to the patient—is an expense to be saved whenever there is a crisis in health service finance.

With the British hospital market put on ice but backed by the general acceptance of the comfort and relief obtainable by the heating film, we had to conclude that the appeal had to be made essentially directly to the actual or potential sufferers. They would constitute the mass consumer market and selling to them would require similar methods and means as to other mass consumer markets. The problem was therefore similar but less formidable than for the food project and amenable to attack on its fringes, by a relatively modest attempt. And we were successful in organising such an attempt.

In the 1960s among our first products, was a foam filled heating pad, soft and flexible to follow contoured shapes. Among our early customers were the supply houses for professional photographers who used the low voltage pads in their dark rooms for keeping development dishes at the desired temperature. Other "private" clients used the pads for propagator units, seed boxes, Jiffy strips, and some friends of my assistants used them instead of hot water bottles. It was this type of product rather than sheets, bandages, plasters, or the like which I had in mind for a first infiltration into the field of medical applications.

As far as expansion in other directions was concerned there had never been any doubt in our minds that America was the largest market in the world for our products, especially for the foil heating film. Indeed Hotfoil Limited, our licensee for sewn heating tapes, sold a sizeable part of its output to America. But the story of our involvement across the Atlantic dated back to before the success of Hotfoil.

Somehow ICI had heard about the project and rung for an appointment to discuss it. I was delighted, and when a high-ranking official came to the laboratory we gave him a full demonstration, and data—indeed all he could ask for.[9] He was obviously impressed and interested. At the time ICI had a 50/50 partnership with the Aluminium Corporation of America, ALCOA, for dealing in this country with aluminium in any form. The two companies jointly owned a factory in Barking, near London, where aluminium foil rolling for the packaging industry was carried out on a large scale. The fact that our foil heating film for heatable foodpacks was an aluminium foil film and that its success promised to enlarge the market for aluminium foil considerably, appeared to encourage the interest of ICI/ALCOA.

The next news from ICI was that ALCOA had the sole rights with projects like ours so I was invited to Pittsburg to demonstrate the project and start negotiations with ALCOA directly. I naturally agreed and my wife and I flew out to Pittsburg, where we stayed with Walter Blenko. Blenko was the principal patent lawyer of the American Technograph company, who also undertook to be my legal representative vis-a-vis ALCOA. He knew most of the executives and lawyers of ALCOA through his many patent litigation cases for and against ALCOA and based on his experience he insisted that all written communications between ALCOA and myself had to go through him.

I had brought with me a full bag of working samples of Hotpacks, a dozen or more different types of single portion packs in duplicate or triplicate, a 120/12 volt transformer, clips and a pile of notes. The demonstrations went like a dream. The audience was visibly and audibly impressed and pelted me with questions and suggestions. I could not have wished for a more enthusiastic reception.

The following day a meeting was arranged with two directors of ALCOA and at that meeting the terms of agreement were discussed and initialled. I left all the samples with one of the technical managers who intended to forward them to one of their laboratories. We felt very encouraged and had a celebration dinner in a restaurant overlooking a brilliantly lit Pittsburg. We then flew back to London feeling that the incredible had happened: we *could* get a very big organisation to accept an important invention from an outsider.

Back in London I thought that I could now leave the Food Project in the hands of ALCOA. However, a few weeks after my return from Pittsburg a request for more samples arrived; they were sent off politely and quickly. A few weeks later came the same request; again I complied without delay. When the third request arrived I became curious.

It transpired that the samples went to a metallurgical laboratory a few hundred miles from Pittsburg where they were given to a junior assistant who had no idea about even the most primitive electrical relations and had connected all the samples to the mains so that they had burst into flames as soon as the current was switched on.

Having found that out I sent new samples with a long and polite letter explaining again how they should be connected up. The letter was understandable, I was sure, to any child. The reply I received was stranger still: they had not been able to find a transformer weighing less than a ton, and the implications of this for portable heating fouls was not encouraging. At that point I too began to have doubts, not about the project, but about the reliability of ALCOA.

I arranged for the local Westinghouse representative to visit the ALCOA laboratory and offer them a small 120/12 volt transformer and I wrote another letter trying to convey to them the simplicity of Ohm's law and reiterating how to connect up the samples. I also sent the final set of samples we had in our laboratory. Their response confirmed that they now had a suitable transformer but that they wanted yet more samples for which they would pay only a tiny amount, calculated from the costs of a production run of ten million. I had had enough and I did what a giant like ALCOA probably never expected: I threw the whole agreement back.[10]

After several other prolonged but finally unsuccessful dealings with very major American corporations we took another line. We designed, experimented with, developed, and finally prototyped certain low cost, basically novel products in which the foil heating film was the main and essential part.

The first types of product we chose were motor accessories to be connected to the power supply by inserting a plug into the cigar lighter socket provided in every American car. There were more than a hundred million cars, as well as many commercial vehicles in America providing unquestionably a colosal potential market for our range of heated appliances.

It was clear to us that we could not get any of these products completely produced in Britain and then try to export them to America. The idea was to export the foil heating film component and license one or more companies there to make and sell the finished product.

From the knowledge I had acquired during and since my printed circuits days in America I was quite certain that there would be no problem in getting the products manufactured in the States. Selling was, on the other hand, quite a different problem. The public needed to be made aware of new products and their qualities, and all means of reaching them, from television and the press, to exhibitions, mail, or stunts, were very expensive.

The stories of two such licensees in America prove revealing. The first concerned the late Dr Eric Kluger, who had come to London from Detroit for an option on the Foil Battery.[11] Our contact had then given him some insight into various aspects of the whole foil heating film project and its potential. When he decided to leave London and return to New York, (where he had no job or business opportunity,) he asked me for a licence on certain products. He was a true gentleman and I greatly appreciated his character, intelligence, and ability. I gave him generous terms and we assured him of the fullest support in all the work our laboratory could do. After he had settled down in Manhattan he founded the firm Heatfoil Products Inc. and made an arrangement with his, and subsequently our, good friend Sal Cenatiempo for organising demonstrations of the products he offered. He knew several influential people in New York and he thought that through them, through letters, visits, and telephone calls he would be able to persuade firms to give him orders for his heatfoil products. Apart from one friend who invested in his firm, he carried the whole financial burden alone. We, in London, gave all his requests top priority, not necessarily for business reasons. For myself it was a prolonged and costly act, but we were impressed by his courage and endurance. I should like to record here just two of the many episodes of his enterprise, as they are not atypical of what can happen to outsiders plugging a novelty in the modern business jungle.

From the beginning of his venture, Kluger was fascinated by the overwhelming growth of fast food chain stores. Their product, in a motorist's country like America cried out for our hotpacks. We all believed that, and he set out to convince the chains. He worked his way to various top managements, demonstrated the packs which we in London had sent him specially for each fast food, and submitted scores of samples to their R & D departments. In the end he received vague approvals for the packs from all the chains, but only one was prepared to consider them straight away, the Pizza Inn. One was, however, a start and Kluger was triumphant. He obtained plugs and clips from Taiwan and we sent a large quantity of foil heating films on paper backing, designed and developed by Vic, to be inserted in the box before the hot pizza was placed inside. The pizza would then lie on top of the insert and be kept hot for as long as desired. Kluger had argued his case as follows: 48 percent of the customers of the average Pizza Inn were motorists, who took the hot pizza away in their car. While in the car the pizza cooled down and after about fifteen minutes had lost its flavour. As a result the number of motorists who bought pizzas from a particular Inn was limited to those who had only about three miles to drive before eating. By incorporating the insert it would be possible to keep the pizza hot indefinitely, immediately increasing enormously the catchment area for any Pizza Inn.

The head office of the chain, in Dallas, accepted Kluger's proposition and ordered a market test. Then, for reasons unspecified, the test was done during a heatwave in July in Springfield, Missouri, at Inns with practically no motorist customers. Local television and other advertisements came out a month too early and concentrated on the plug and clips, costing $1.29, instead of the 10 cent insert with the Hotpack. Sales girls had no knowledge of the affair and only one Inn manager knew anything about it. Under these circumstances the test was, of course, a complete flop. When the facts of this breakdown of middle management became known to the head office they sent a letter of apology to Kluger, but that was no consolation for all the lost time and effort.

A very different story unfolded with another of our licensees, in California. Arnold Fehl, a younger cousin of mine, (one of my few relatives to escape the holocaust,) had settled near San Fransisco. After he returned from the war, (when he served with the American forces in Germany,) he studied electrical engineering in Chicago. He found a job in an electrical switchgear firm and became an expert in that industry. Eventually he went West and started his own firm Sierra Switchboard, which grew to become the largest independent manufacturer of electrical distribution equipment in California.

In 1975 my wife and I took up the repeated invitation to visit Arnold, his wife and his two sons, and we stayed with them for two weeks. The younger son was in his last year of law study and the elder son, Richard, had just returned from Los Angeles and was for the time being working at Sierra Switchboard. Our visit gave the opportunity to discuss their desire to diversify, and thereby to provide a chance for Richard to show what he could do. Arnold would be prepared to help with finance and general support from Sierra Switchboard. The outcome of the discussion—which went from the analysis of the psychological and the personal to financial, organisational, marketing, and technical matters—was the decision to found a new company for the production of certain foil heating film products which I and Vic had fully developed, and for which they thought there was a market in California.

Richard was to be in charge of the organisation. I suggested the name: Thermal Technology Inc., which was unanimously acclaimed. As to the products the enterprise should launch they decided after long debate to start with waterbed heaters before going on to a more ambitious project. The choice of waterbed heaters—a fast growing market on the West Coast—was governed by several factors. They had personal friends in that industry who would join them as engineers and in marketing; we, in London, had developed and produced heating film for medical waterbeds and could give them all the technical knowhow required; the waterbed project was a relatively new, simple project which did not need much investment; and finally as the first project it would create an organisational nucleus for tackling the next, larger project.

I gave them a licence on my patents, shared our knowhow for waterbed heaters, and agreed terms for the supply of the foil heating film. Arnold even had an option to buy a machine from us on certain terms. I treated the project as a real family affair, an exception to the general rule, not to do business with members of the family, as I believed it did not apply in my case. I could clearly never use the law against a relative, but as I would not use it against any licensee anyhow, what difference did that make: We flew back to England and the job of giving birth to Thermal Technology began in Menlo Park, California, where Arnold had his factory.[12]

In the next phase Vic produced innumerable samples for them, as well as prototypes, technical sketches, and letters to guide the young enterprise through often stormy waves of despair, disappointments, mistakes, anxiety, and inexperience. Their first crisis looked quite serious and Vic and his wife went over to stay with them. He succeeded easily in sorting things out and putting the organisation back on an even keel. He made good friends with all the Fehls and the staff, and since then Thermal Technology's production and sale of waterbed heaters has become well established, growing year by year, and the company had become quite profitable by 1980.

The example of the two licensees, shows on the one hand the successful launching of a novel product by a single organisation with its own apparatus of production and sales. On the other hand, the launching of new products by an enterprise having to rely on a chain of independent parties over which it has no real control led to failure. The difference illustrated is often referred to as one between a vertical and horizontal organisation. We were fully conscious of the fact that even we, as licensors, had no sales organisation; that is, we were only half a vertical set up, which limited our growth. But with us that was, and is, a deliberate policy; we wish to remain small and manageable rather than becoming a bureaucracy no matter how wealthy. Licensees, however, usually want to grow, and we have to support them without being infected by their aims.

10: To Be or Not to Be an Inventor?

In Vienna in the 1930s my need for a way of escape from the Nazis led me to paraphrase Hamlet's famous question, "to be or not to be an inventor", thus turning his problem from one of philosophy to one of a very practical nature. My positive answer to the rephrased question led to a patent application which in turn got me an invitation from Marconi who were interested in my very academic invention of stereoscopic television with a small frequency band. This invitation permitted me to come to England and thus probably saved my life.

My patent application was, then, a peculiar key to get me into this land of hope and glory. Obviously that was not the way in which patents were usually used.

Many students of science and many technologists consider the advantages and disadvantages of the life of an inventor put such questions to themselves whenever they think of commercially implementing a novel idea. Every year tens of thousands of patents are applied for in England alone, many more in the United States and around the world; a whole profession of patent agents exists and many lawyers earn their livings by dealing exclusively with patents. A decision to earn a living by being an inventor, therefore, means coming to terms with this system.

What then makes an inventor? Is he or she just someone who has a certain peculiar psychological characteristic inherited or acquired? Or is he a person who becomes bothered by one technical problem he finds himself unable to leave alone? Whichever he is, if he feels the urge or duty to solve particular problems or simply occupies his time thinking up new schemes, once a particular solution to a technical problem occurs to him he has made the first step towards becoming an inventor. If he is to take this any further he is then faced with the need to apply for a patent.

A patent, particularly an American patent, is granted only after a thorough investigation of freedom from anticipation and it defines the invention clearly in legal language. It is subsequently free of annual renewal fees and is certainly, in spite of being part of a cumbersome system, a valuable, often indispensible help for the inventor. At least that is what it is under the most favourable conditions. However, when the patent is granted what the patentee obtains is the *legal* right of a monopoly in exploiting his invention for a given number of years. There are two problems with this: a monopoly situation is not necessarily what the patentee wants, and as it is the law which grants these rights it is through the law that they must be defended.

From the walls of the Patent Office in Washington Abraham Lincoln tells us, "The patent system added the fuel of interest to the fire of genius." Does it do that for the independent inventor or for the small organisation?

A patent gives to the patentee exclusive rights formulated in the patent claims. These rights can be strong, sometimes decisive arguments in sueing an infringer. Strong claims must however be backed by a financially strong organisation interested in the invention, and prepared to fight a court case against an important infringer. Legal expenses are very high and court cases can take a very long time. The freelance inventor's exclusive patent rights are of little value to him for such inventions which he cannot bring to a marketable stage himself. If he is lucky he can perhaps find a large firm interested in the invention and somebody of power within it to guide the innovation through all the technical and bureaucratic impediments it will encounter. But such luck is rare indeed. In other words the patent system alone may not be enough to persuade inventive individuals to carry on inventing.

Nevertheless my attitude is not overall anti-patent. Since 1957 I have faced Hamlet's amended question again and again. Each time I answered it positively, and I built up Foil Engineering Limited as my production unit. While doing this I did not rely on patent rights only, but I took advantage of them in some ways. Indeed, I liked and used aspects of the patent system. The formulation of a patent application required very precise thinking to identify the new and practical nucleus of an idea. That was a rewarding activity, which often generated new ideas. I saw the whole process as a link in the chain of inventions on which technological development is based.

This did not prevent me from dreaming about how much better the patent system might be if patent rights were not granted for a monopoly, but if, instead, their free use were linked to facilities for the inventor to develop his invention and to obtain reasonable financial recompense. This could be quite modest and financed by a differential tax, maybe along the lines of Value Added Tax, levelled on competing products which do not use the invention. These have been no more than dreams, hopeless of realisation at the present in the face of the vested interests of large corporations and, no less, the legal profession. But how many of the reforms in our system have started as dreams?

Looking back on the playground of the patent system with the inventor's dreams on one side, and the intrigues, fights, and politics of large organisations on the other, we can restore ourselves to calm with the thought that the dangers in the system, so far experienced, are at least tolerable. The patent laws, we could be confident, dealt only with inanimate products, methods and means. But there has been a significant, and possibly disturbing event in patent law recently, on which Professor Leon R. Kass wrote in *Commentary* (72:6, December 1981), under the heading "Patenting life".

On 16 June 1980 the Supreme Court of the United States ruled that a living microorganism was patentable matter, under the provision of patent laws enacted by Congress in 1952. In 1972 Ananda Chakrabarty, a microbiologist at the University of Illinois, had filed patent application; assigned to the General Electric Company, asserting multiple claims related to a novel bacteria strain that he had obtained with the aid of techniques of genetic engineering, a strain capable of degrading many components of crude oil and thus potentially useful in the biological control of oil spills. In addition to readily granted process claims for the method of producing the bacaterium, and claims relating to the mode of carrying such bacteria to water-borne oil spills, Chakrabarty claimed patent right to the bacteria themselves. This last claim, at first rejected by the patent examiner and then by the Patent Office Board of Appeals was finally granted on appeal by the United States Court of Customs and Patent Appeals in 1979, in a decision affirmed by a narrow five-four vote of the Supreme Court a year later (Diamond v Chakrabarty 447 U. S. 303).

Patent claims are now pending for other living microorganisms as well as for animal cell lines propagated in tissue culture, allegedly valuable for uses ranging from a cheaper means of making penicillin to novel treatments for specific cancers. Genetic-engineering firms are springing up all around. Academic molecular biologists are being courted by industry, with astounding financial incentives. Major grants for genetic-engineering research to universities have been given by industries in exchange for patent rights to any resulting useful and profitable discoveries.

Patenting life is perhaps the deepest philosophical problem facing scientists concerned with genetic engineering. Furthermore, a sincere answer to the question, to be or not to be an inventor, has become even more complicated ever since scientists working on the atomic bomb have alerted mankind to the danger of where inventions in pure science may eventually lead. Maybe fortunately for me, my inventions were in the applied sciences field, in corners of that field which were distant from the applications of nuclear physics and genetic engineering.

How to earn a living was one of my major problems, and one which stayed with me for many years, and to which I had to return time and again. Until comparatively recently I had no security; but I did not miss it. Life was at times quite hazardous, but I never once chose to give up my activities as an inventor.

So why did I go on inventing as a freelancer? Perhaps one reason was the intense reaction I always felt against inefficiency: I deeply resented inefficient methods and means. There were also accidental reasons: having been prevented by partial colour blindness from studying medicine I had become an engineer, whose job I believed it was to contribute to an improvement in the real life economy of society. This "life economy", according to L. L. Brown, the head of a firm who showed interest in one of our products, is "the study and comparative measurement of the efficiency with which natural resources are used, converted and distributed by human beings who are the most important natural resources of all." The vivid experience of the gross inefficiency with which all these functions are carried out in any big firm would probably have driven me out of any post in such a firm. Within them Parkinson's Law rules and the politics of power groups occupy the mind. However much I might sometimes have wished for a regular job, it would never really have suited me. It seems that it was my fate to remain a freelancer.

My wishes and aims were to produce useful items without having to sell them myself. This book shows how I tried, and to what extent I achieved my goal. I can certainly claim two major inventions, the first of which, the printed circuit, will stay in my heart as my contribution to Hitler's defeat. The proximity fuse was a decisive weapon against Hitler's V1 rockets, the "Doodlebugs". But the printed circuit turned out to be more than just a clever idea. After the war, printed circuit technology became the basis of production within the whole electronics industry the world over, and has remained so until the present day, already more than forty years on. Its principles of design have given birth to the transistor, the integrated circuit, the "chip", and the microprocessor. It has proved to be a great invention. My other major invention, the foil heating film, created a means for surface heating, a basic technology, of which only the first applications have so far been realised.

Both these inventions fill me with pride and go a long way towards compensating me for the numerous occasions when I had to watch ideas sink into a swamp of bureaucracy, for times when I was unable to arouse sufficient interest, and for times when the patent system failed me.

The foil battery was a typical case where, as a freelance inventor, I had no other way open to me but to try selling the idea. Large investment would be required for initial practical tests and further exploitation. But no big firm took up the foil battery in spite of its merits. Moreover, as far as I could tell, it was not my failure to carry conviction: management in the whole battery and car industry, as well as the R & D people, regarded it as some sort of time bomb. They feared that their established but inefficient technology would be unable to withstand the competition. In this case my patents, however basic strong, did not help.

I should like to conclude this book by citing two notes of a more optimistic nature about inventions. The first is a remark made by the great physicist Max Planck, commenting on new ideas:

> An important scientific innovation rarely makes its way by gradually winning over and converting its opponents; it rarely happens that Saul becomes Paul. What does happen is that its opponents die out and that the growing generation is familiarised with the idea from the beginning.

The second is a quote from Albert Einstein, "I know why there are so many people who love chopping wood. In this activity one immediately sees the results." If I may paraphrase this aphorism "I know why there are so few people who make a basic invention. In this activity it takes a long time to see results."

The more fortunate the inventor whose activity has borne fruit in a relatively short span of time, for my invention has now indeed come of age.

Today my technology extends far beyond the strict electronic industry and there is hardly any aspect of our electrical, mechanical and chemical engineering, of our communication and transport and of our science and cultural life, in which printed circuits have not made a contribution. And this contribution I have seen cultivated, extended, and developed by an energetic resourceful, young generation of technologists while I myself could turn to new areas of invention.

An inventor, seeing his expectations met to such a degree, is indeed fortunate. When he also receives recognition by the highest legal authorities, the High Court, Court of Appeals and the House of Lords, it seems that there are, with not too extended a time scale, other ways for producing visible results than chopping wood.

List of My Printed Circuit Patents

UK

List of my British Patents on Printed Circuits assigned to Technograph Limited (all having corresponding US Patents and some in other countries)

Patent	Year	Title
245 43	1936	Printed Circuits
639 111	1943	Three-dimensional Printed Circuits
639 178	1943	Foil Technique of Printed Circuits
639 179	1943	Powder Printing
644 565	1945	Production Windings
639 658	1947	Tin Printing
685 912	1947	Concertina Folding and Miniaturisation
700 451	1947	Slicing Technique
690 328	1948	Printed Circuit Capacitors
690 360	1948	Dielectric Materials
700 452	1948	Telephone Exchange Equipment
700 496	1948	Sliced Components
670 926	1948	Electric Fuses
690 691	1949	Multilayer Materials for Printed Circuits
690 696	1949	Production of Coated Film
700 458	1949	Infinite Inductances
695 386	1950	Metallised Insulators
700 459	1950	Tape Cable
700 490	1950	Connections to Foil Conductors
728 219	1951	Adhesive Layers
732 437	1951	Grid-like Components
742 251	1952	Electric Heating Mantles
710 235	1952	Electric Spark Method of Printed Circuits
728 606	1952	Treatment of Foil Resist
746 936	1952	Stepping Stone Circuit Patterns
764 134	1952	Variable Capacitors
764 017	1952	Variable Capacitors
792 145	1953	Mechanical Movement through Electric Heating
835 214	1955	Anisotropy

US

List of my US Patents, pertaining to Printed Circuits and Components, offered for licence by Technograph Inc.

Patent	Date	Title
2 582 685	15 Jan. 1952	Method of Producing Electrical Components
2 602 731	8 July 1952	Method of Making Circuit Panels
2 607 825	19 Aug. 1952	Electric Capacitor and Method of Making it

2 634 310	7 April 1953	Electrical Connecting Strip
2 662 957	15 Dec. 1953	Electric Resistor or Semiconductor
2 666 254	19 Jan. 1954	Electrical Windings
2 703 854	8 March 1955	Electrical Coil
2 706 697	19 April 1955	Manufacture of Electric Circuit Components
2 729 884	10 Jan. 1956	Metal Article Deforming Method and Apparatus therefor
2 736 677	28 Feb. 1956	Metallised Insulators
2 737 571	6 March 1956	Electric Resistance Heating Device
2 747 977	29 May 1956	Method of Making Printed Circuits
2 748 071	29 May 1956	Apparatus for Regeneration of Etching Media
2 758 256	7 Aug. 1956	Electric Circuit Components
2 778 762	22 Jan. 1957	Electric Capacitor and Method of Making it
2 785 280	12 March 1957	Printed Electric Circuits and Components
2 789 259	16 April 1957	Variable Capacitors
2 802 995	13 Aug. 1957	Printed Circuit Connection and Method of Making Same
2 874 360	17 Feb. 1959	Electrical Windings
2 885 524	5 May 1959	Electrical Resistance Devices
2 886 880	19 May 1959	Method of Producing Electrical Circuit Components
2 904 761	15 Sept. 1959	Magnetic Circuit Components
2 904 772	15 Sept. 1959	Printed Circuit Construction and Method of Making

Hotfoil and Foil Heating Film Patent Structure

MY UK PATENTS

Patent	Title
900 515	Electric Surface Heating Devices
900 516	Electric Heating Resistance Strips
900 517	Surface Heating Device
900 518	Electrical Heating System
900 519	Electric Conductor Strips
905 867	Manufacture of Electrical and Conducting Heating Devices
908 680	Electrical Space Heating System
912 980	Production of Metallised Laminates
914 952	Electrical Heating Devices
939 292	Method of Heating in Vehicles
973 211	Electric Tape Cables
992 571	Heating Packages
992 572	Improvements relating to Electrical Connectors
992 573	Heating Packages
992 574	Heating Packages
1 020 311	Electrical Heating Film
1 020 312	Electrically Conductive Films
1 020 313	Producing Changes of Physical State by Electric Heating
1 020 314	Heating of Material used in Building, Civil Engineering, etc.
1 020 315	Heating Electrically Conductive Films in Movement
1 020 316	Electric Heating Film
1 020 317	Electric Heating Means

1 020 318	Displacement of Materials with the Aid of Electric Heating
1 115 642	Space Conditioning Systems
1 115 643	Multilayer Wall Covering for Space Heating
1 115 644	Connections for supplying Electric Current to Heating Film
1 132 965	Improvements in the Manufacture of Slits in a Web
1 132 966	Electric Heating Film
1 155 81	Improvements in Packaging
1 155 82	Improvements in Packaging
1 176 651	Electric Heat Treatment of Structural Materials
1 187 562	Improvements in Electric Heating Devices
1 192 13	Electric Heating
1 193 621	Selective Heating
1 193 622	Controlled Heating
1 193 623	Controlled Heating
1 234 445	Washing Unit
1 333 049	Improvements in and relating to Heating Panels

MY US PATENTS

Patent	*Title*
2 971 073	Electrical Surface Heating Devices
3 020 378	Electrical Heating and Conducting Devices
3 026 234	Laminates Embodying Electrically Conductive Patterns
3 033 970	Electric Conductor Strips
3 089 017	Electric Heating System
3 099 540	Electric Foil Resistance Dryer
3 100 711	Food Package
3 132 228	Method of Heating in Vehicles
3 149 406	Method of Making Electrical Heating and Conducting Devices
3 283 284	Electrical Heating Film
3 296 415	Electrically Heated Dispensable Container
3 372 487	Method of Drying by Electrical Means
3 317 657	Flat Electric Cables
3 408 735	Method of Making Patterned Foil Webs
3 410 336	Thermal Conditioning System for an Enclosed Space
3 413 439	Electrical Circuit Connections
3 473 003	Wall Covering Material for Use in Space Heating
3 483 358	Electrically Heatable Package
3 510 547	Method of Heat Treating a Body of Curable Material
3 516 218	Packaging Method
3 522 415	Electric Heating Devices
3 523 542	Hair Curling and Straightening Means
3 539 767	Space Heater having Electrical Resistance Heating Film
3 539 768	Electrical Space Heating System
3 539 892	Heatable Package with Displaceable Fluent Substance
3 544 762	Method Separating Portions of a Transported Body by Resistance Heating
3 546 432	Wall Covering Material for Use in Space Heating
3 567 353	Thermal Conditioning System
3 573 430	Surface Heating Device
3 575 027	Manufacture of Patterned Webs
3 751 629	Surface Heating Device

3 721 800   Electrically Heated Package
3 736 404   Combined Demisting and Defrosting Heating Panel for Windows and other
            Transparent Areas
3 829 654   Electrically Heated Package
3 846 204   Heating Methods
3 897 928   Concrete Mould

Appendix 1: Uses of the Heating Film in Slicing and Preserving

Most ideas in this appendix deal with the use of the heating film in fields not usually within the realm of heating technology. There exist hardly any reference to them in print other than my patents. Fuller information and more drawings are available from the patent specifications.

The Heating Film: A Parting Tool

It may be interesting to start with some ideas whereby the film is not applied to heat food, but rather for some other purpose.

The first example of this type of application is the use of the film for the subdivision of a large frozen body into smaller slices. A problem of such sort is usually dealt with by mechanical forces using familiar cutting tools. I had the idea to solve it by incorporating a foil heating film at all planes at which the body is to be subdivided. The cohesion of the parts of the body across the heating film will be strong enough to keep the whole body intact at ordinary temperatures. But a heavy heat shock over the whole area of the film will suddenly reduce the cohesion of a thin layer on or around the film so that the body is or can be easily parted then and there. This heat-slicing can be done slice by slice in succession, the slices sliding off the bulk body or altogether by switching on all heating films simultaneously.

A great variety of goods and substances can be solidified or allowed to solidify into a solid in which heating films are interspersed. These films cover such cross-sectional areas and are spaced at such distances that the slices or sub-blocks into which the bulk block will be partitioned are of the shape and weight desired at the distribution point. This partitioning is effected almost without effort when the interface between the solid and the heating film becomes hot. The interface may melt. If it is arranged that the film gets very hot for a few seconds only, the layer may stick on the film and not stay on the slice. Frozen fish is one illustrative example.

The slices or sub-blocks of frozen or solidified material between which the heating film is interleaved and which cohere in a bulk block need not be slabs with large flat and parallel faces, but can have any shape which permits them to interlock with, or nest and stack one in another provided that the shape of the mating faces permits adjacent slabs to slide over one another or to be separated and become free by a very small relative displacement as soon as the interface is brought to the melting point.

Instead of coating packages with hot melt adhesive, a heating film with a layer of hot melt adhesive or with a paper impregnated with a hot melt adhesive on both sides of the metal foil pattern can be used and the packages be stuck together by placing this type of heating film between them and keeping them pressed together while the film is connected to a low voltage supply and afterwards until the adhesive has cooled down again. (For further information and sketches see BP 1,192,013 or US, 3544762.)

Independence from the Autoclave: A Revolution in Food Preserving

An important invention of mine (UK Patent 1,155081, US Patent 3,516 218) relates to the hermetic packaging of goods needing to be heat-processed at the time of packaging. Examples are foodstuffs which need to be sterilized and cooked; pharmaceutical products; chemicals which may need to react at the time of packaging; and some surgical instruments needing to be sterilized at the time of packaging.

In all such cases, if the package is hermetically sealed before heat treatment, there almost invariably arises the problem of pressure differences developing within the pack during the heating. To take the typical example of foodstuffs which generally contain water: to sterilize as well as cook them it is necessary to raise the temperature to say 240°F (116°C) at which the vapour pressure of water is well above atmospheric pressure. A common method is to effect the heat treatment with steam at the pressure corresponding to the required temperature. For the above temperature and saturated steam this is about 25 lbs. per sq. inch (1.7 atmospheres) in value, and the operation needs to be done in an autoclave. Moreover to ensure that the goods are brought to the required temperature in a reasonable time, steam must be at a high temperature and therefore a higher pressure must be used. At the start when the package is cold, pressure within it will be approximately atmospheric while the steam pressure is much higher. This pressure difference remains the same until the temperature within the package reaches boiling point at atmospheric pressure. Thereafter as the temperature rises within the package so does the pressure until finally the internal and external pressures balance or nearly balance. If the package is now taken out of the autoclave to cool under ambient conditions, the internal pressure will be much higher. There are thus two phases, heating and cooling, when there is a very substantial pressure difference between the inside and outside of the package and the walls of the package must be strong enough to withstand this. Similar conditions will usually arise with the other types of package referred to.

My invention avoids substantial pressure difference between the inside and outside of the container by effecting direct heating of the container while leaving it unsealed. Merely to leave the container open would not be a satisfactory solution, if steam or hot air heating was employed, as this would cause condensation or drying or other damage to the contents.

The invention proposes a simple method of sterilising and/or pressure cooking food and the preservation of this food in any package which stands the sterilising temperature of, usually, 240°F (116°C) and which keeps the food alright in storage. Its central feature is to keep the container of the food open while it is placed in a compressed air chamber and the food in the container is directly heated to just below the boiling temperature at the high air pressure in the chamber. This boiling temperature is higher than the sterilising temperature. As the wall of the open container is not exposed to stress during the heating or cooling of the food in the chamber and hermetic sealing of the package takes place only after the food has cooled down to below 212°F (100°C), it is possible to replace the tinned steel can by a plastic pouch or aluminium foil and aluminium foil laminates.

The direct heating of the container and the contents is done preferably by heating films incorporated in the package or contacting the package over as large a surface as practical.

The fact that the present invention uses direct heating of the packages and compressed air which does not get too hot in the process, permits such air pressure vessels and batteries of such vessels side by side to be used in the farmer's field and makes canning and sterlising a much less cumbersome, a cheaper, a better, and more controllable process. The equipment can, for example, be mounted on a truck to render it mobile and thus enable it to be used in the packaging of sensitive products such as various fruits and vegetables, when they are not merely fresh, but almost immediately after they have been harvested.

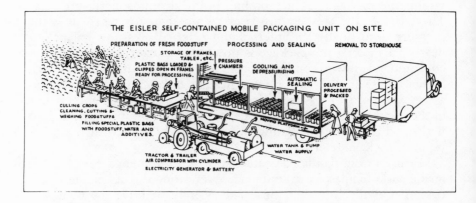

The proposed process, which is illustrated in the accompanying sketch is in principle quite conventional, except that it is done in the field using quite simple means, without an autoclave. It eliminates trade intermediaries and preserves a maximum of the vitamin content and freshness of the food.

A van would pull up by the side of a field. There a crop is gathered, washed, cut up, weighed, and filled into packs together with a measured amount of water and additives. The packs are left open and go into a pressure chamber on the van, which is supplied with clean air at high pressure from an air compressor on a tractor-driven trailer. While under that air pressure, the food in the open packs is "air-pressure-cooked" (without raising steam) to sterilisation temperature (240°F) by foil heating films with which the packs are equipped. The electric current for this heating is supplied by a battery and generator which is also driven by the engine of the tractor.

After having passed the required time at sterilisation temperature, the food in the open packs is allowed to cool while still under pressure down to below boiling point (212°F). Then the packs are pushed into the next compartment under atmospheric pressure, are hermetically sealed, and are then loaded onto a small van to be stored and sold by the farmer at a time of his choice and as a product commanding a much higher price, even if sold much cheaper than the price of the lower quality food of the canneries. And that should leave him with considerably more money as profit, without the risks of perishable goods on his hands, while offering the public cheaper and better, more vitamin containing food.

Appendix 2: Uses of the Heating Film in Concrete and Building

We looked briefly into the possible advantages and difficulties of improved heating of surfaces by using dispensible Heating Film for the following:

Drying damp structures.
Pre-heating surfaces prior to application of hot or viscous compounds to dry surfaces and/or enable easier flow of caulking-, gap-filling and jointing compounds.
Dispensing asphalt and bituminous melts from heatable bags and other heatable containers.
Temporary shelters in winter and general frost-fighting use.

We investigated the Heating Film's use in improved de-icing and frost prevention schemes, by embedding it in the surface layers of roads, airstrips, rail junctions, sports grounds and then through cyclic heating of sections of the surface layer, thereby only melting the interface, and subsequently mechanically removing ice or snow.

Our main interest, however, centered on the accelerated curing of concrete for which we made extensive experiments.

The advance which I thought the heating film could bring to this method would be:

The uniformity of its heating emission which avoids the great dangers to the structure of concrete involved in other methods which space hot wires on the concrete surface or withiin the mass of concrete. Such inhomogeneous heating may cause cracks and stresses in the concrete and it limits the permissible time cycle.
The convenience and labour saving when placing heating film elements on or inside concrete moulds.
The freedom from danger and from insulation troubles. No electrical knowledge required.
The low cost of the heating film and of the accessories required.
The freedom of choice of the most desirable time cycle.

141

Our stand at the International Precast Concrete Exhibition, London, 1965.

The almost complete mastery of the curing behaviour of the concrete, predictable in the design office, when fixing the position and loading of the heating film for any particular structure.

We investigated on site wiring and precast wiring. Further information is available in our patents.

Appendix 3: The Heating Film and the Chemical Industry

We did not touch the food projects allied applications after the break with Alcoa although the ICI itself had shown interest in using the heating film with one of its own products after we had carried out a highly successful experiment. That allied project was the packaging of chemicals in electrically heatable dispensible containers. The experiment we carried out was the heating of a 50 gallon drum of Phenol, ICI being its major supplier to industry. Phenol has a melting point of 47°C.

So far the position had been as follows. A large proportion of ICI's production was shipped in steel drums. The hot liquid phenol was filled into the drum and solidified as it cooled to environmental temperature. The customer, usually a chemical manufacturer, placed the drum into very hot water and poured out the heated liquid stuff. If he only wanted a part he had to repeat the procedure and finally dispense with the steel drum. The whole dirty business was wasteful and expensive.

We supplied a heating film on a simple polythene base, had it placed in a fibre drum before the hot liquid phenol was filled in. No other change of procedure was necessary. In order to liquify the whole or any horizontal slice of the solid phenol the whole or only a part of the heating film was energized. We used 3000 watts per sqft! We could go to such a high rating because the film was completely immersed in phenol and never reached much over 50°C, so there was no danger of gas bubbles. The advantages opened to ICI and the customer were self-evident: light, easily disposable fibre drums, instead of steel drums, cost less as container and in transport. Emptying them would be cleaner, quicker, and cheaper. The main disadvantage would appear to be the need of a large transformer. Its cost should soon be amortized by the saving of labour and fuel even for a small chemical manufacturer.

I do not know why ICI did not come back to us after the success of the experiment which their own people acclaimed. It was possible

144

their agreement with Alcoa—apart from the always present bureaucratic ramifications in such a gigantic organisation as ICI. Packages using aluminium foil were—as we understood at that time—Alcoa's monopoly and our break with Alcoa may have put an end also to ICI's possible interest in their use for its own chemicals.

That scheme, however, remains one of my "grand schemes" although we have not promoted it anymore except for the accompanying write-up which we made for Avery and Kanthal as an example of what the future could hold. As far as I know they kept it in their files and did not even distribute it to potentially interested chemical enterprises.

We considered the development of an electric heating film of such a negligible cost that it can be thrown away after a single use. This would permit the safe, convenient, and rapid heating of substances in their non-returnable containers.

The chemicals that can be dispensed hot from their container include not only normal liquids of practically any type but also highly viscous and solid materials such as asphalt and other bitumenous compounds, heavy oils, waxes, solventless paints and inks, hot-melt adhesives, thermoplastics, some elastomeric mixtures and thermosetting resins.

The heating film can be used for the following purposes:

If chemical is liquid and is to be used hot. Can apply to all liquids.

If chemical is solid or of high viscosity and is to be molten or its viscosity is to be lowered for ease of dispensing or application. Examples: coal tar, wax, glue, printing ink, viscous oil.

If reaction or cure is to be accelerated at place of use. Examples: low pressure thermosetting resins, vulcanisable elastomeric compounds.

If special substances are used in compound, e.g. stronger catalysts and accelerators than permissible with mixing vessels in mills. Examples: as above.

If vapour or gas is to be produced from liquid or solid chemical, e.g. as means of pressure and expansion or provision of an inert atmosphere.

In order to achieve savings in solvents or thinners, packaging transport, convenience in dispensing and/or application, saving in user's equipment, fuel, time or labour. Examples: solventless paints and inks, adhesives, thermoplastics, bituminous compounds.

The heating film can be used in the packaging of chemicals. Coal tar, phenol, wax, blue and all substances which are filled into containers in liquid form and solidify: Viscous oil, printing ink, pastes and other substances which are messy and difficult to dispense quickly and completely; thermoplastics, hot melt adhesives and impregnating compounds; Solvent-less paints and inks; two-part packs of neoprene.

Appendix 4: The Foil Battery

(from *Electric Vehicle Developments* December 1980)

There are many mental images of an electric car: mine is that of a two-passenger urban commuter car, extremely well sprung and so light that it can be manually lifted. According to recent studies cited by *The Washington Post* there are only one or two passengers in the present conventional motorcar in 92% of trips to and from work, and in 78% of all trips. Some other statistics show that well over 70% of all trips cover less than 20 miles. From these figures it can be concluded that such a lightweight car should have a ready market and, when used in great quantity, should constitute a great, nationally important, saving in petrol and energy.

This is also a logically self-evident concept as it is incompatible with all principles of efficiency to have to carry with you the mass of the modern car, more than ten times as heavy as you and your passenger, when the purpose of the exercise is just to carry you both from A to B.

It would, of course, require a reorientation of thinking about the electric car, away from the fixed idea that only a very high capacity battery can be considered for the 'electric car'; this would necessarily and prohibitively weigh too much unless we had a new type of battery such as the sodium-sulphur battery which will take decades to be fully developed, generally accepted and able to supply immense markets. The materials used in some conventional batteries are not acceptable for other reasons. We, therefore, are left only with the lead-acid battery and one or two alkaline types; but these in their conventional construction weigh too much and require a long time for recharging.

The way out of this dilemma is by using a foil battery of relatively small capacity, and for that reason alone, of low weight. Such a foil battery can be *recharged at mains voltage in a few minutes* and has adequate cooling features. It is this concept, with the emphasis on the extraordinarily rapid recharging possibility, which requires a reorientation of thinking.

Although rapid recharging of an alkaline battery should be easier, I

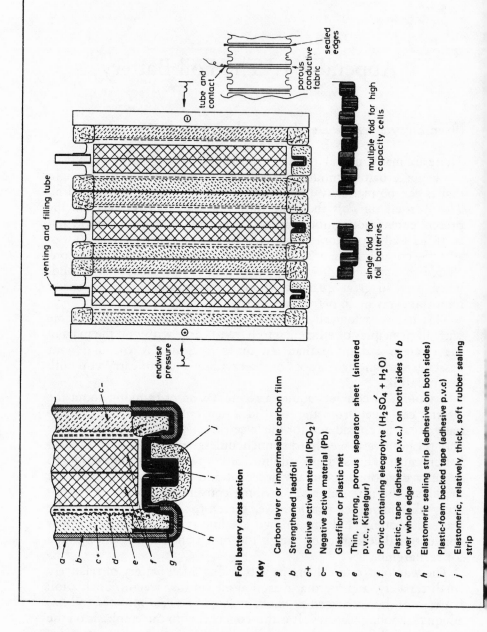

sealed edges

tube and contact

porous conductive fabric

multiple fold for high capacity cells

single fold for foil batteries

venting and filling tube

endwise pressure

Foil battery cross section

Key

a Carbon layer or impermeable carbon film

b Strengthened leadfoil

c+ Positive active material (PbO_2)

c- Negative active material (Pb)

d Glassfibre or plastic net

e Thin, strong, porous separator sheet (sintered p.v.c., Kieselgur)

f Porvic containing elecgrolyte ($H_2SO_4 + H_2O$)

g Plastic, tape (adhesive p.v.c.) on both sides of b over whole edge

h Elastomeric sealing strip (adhesive on both sides)

i Plastic-foam backed tape (adhesive p.v.c)

j Elastomeric, relatively thick, soft rubber sealing strip

favour the lead-acid battery system as a practical proposition for a mass scale use in a relatively short time. We have the longest and most intimate experience with this system for cars, all materials for it are inexpensive, are available in practically unlimited quantity, and knowledge of and confidence in it is widespread over the whole world.

Perhaps I am also biased towards the lead-acid system as the foil battery, its problems and claims, which is the theme of this article and which I invented some twenty years ago* concentrated essentially on the lead-acid type. At that time I was not thinking of its use in the electric car but in various other applications. I proposed to develop a new type of storage battery, and more specifically, a new type of lead-acid battery which uses the same chemical processes and active substances as the conventional battery, but differs from it fundamentally in structure, working and manufacture. The invention on which the proposal was based is applicable to all secondary storage batteries, but it was proposed to start with the development of the lead-acid type because it promised greater financial rewards to the sponsors.

A typical foil battery is made up of self-contained modules which are the production and trading units and have the size and shape of a book held upright. Like a book, the new battery composed of many modules is essentially a pile of thin sheets held pressed between end plates, the sheets of the battery being foils of various conductive materials forming cell walls enclosing gauzes and porous insulating materials filled with electrolyte which cannot flow out because the sides and the bottom of each cell are sealed. Each cell consists of a thin 'barrier' coated with a thin film of active material forming the positive place, a porous separator carrying the acid, and another thin film of active material as a negative place mounted on a further thin 'barrier' as illustrated. The outer faces of these 'barriers' then form part of the neighbouring cells and are again coated with active material forming the place with the opposite charge to that on the reverse side. The series connection between cells is through the 'barrier' which is chemically impervious but electrically conducting. Because the layers are thin, the whole of the active material is effective; in a conventional battery a large part of the material is electrochemically idle for much of the operating time. This is the basis of the high efficiency of the foil construction.

For use in a conventional car, a module of six cells would be made giving a nominal 12V and a number of these modules would be connected in parallel to provide the total electrical capacity needed.

The battery would comprise five internal bipolar electrode assemblies sandwiched between two unipolar electrode assemblies, one of the positive single electrode and the other the negative. The intermediate foils have one face of active material as positive electrode and one as negative.

Each active layer is covered with a porous insulating film of say nylon or p.v.c. gauze. For the complete made-up module the terminals are provided by the barrier layers of the outermost assemblies. Edges of the assemblies are bound with impervious insulating tape and are slipped into the centre fold of a concertina folded tape. These are sealed together to hold in position the centre separators within the cells. The porous separator, which may be plastic, rubber or other alternatives, carries the acid.

The barrier layer, where the properties are so important, may be an impervious carbon—for example carbon scrap from tube material used in atomic energy which has been fired at 2573 K. 'Active materials' for electrodes and the associated electrolyte are naturally chosen according to the battery system being used; first proposals have been for a lead-acid battery.

Problems of cooling at high rates of charge and discharge appear readily soluble and the modular construction is of advantage also in this aspect.

The striking constructional characteristic of the new battery is that there is substantially nothing else in it but these thin layers of relatively enormous surface area and minute thickness. The following basic working features of the battery can be readily understood if this constructional characteristic is always kept in mind.

Access to surface and progress of chemical reaction: The thin active layers are in contact with the electrolyte over their very large surface area to and from which the flow of liquid and gases is unimpeded. A substantially larger proportion of the active mass of the foil type electrode can therefore interact with the electrolyte at any time than is the case with a thick battery plate however spongy its material may be. Apart from the relative areas of the surface in actual contact with the electrolyte it can be visualised that the progress of the chemical reaction in the narrow cracks and pores of a thick battery plate is slow compared with the reaction on a foil surface and the progress through a foil thickness. The foil works primarily by direct contact, the thick plate substantially by diffusion and the latter is a slow process.

The location of the chemical reaction to the freely accessible foil surface will be referred to below in connection with its effect on the life of the battery but here it is seen primarily as the basic factor enabling a large part of the whole capacity of the battery to be ready for almost instantaneous discharge or for allowing a very quick charge. I consider this as perhaps the most important qualification for the light electric car.

Safety by low current density: The localisation of the chemical reaction to the foil surfaces does not mean a high intensity of the reaction per unit foil surface area. This intensity of chemical reaction is proportional to the current per unit foil area, that is to the current density. The design of the battery provides as one of its fundamental characteristics that the electric current always flows through the thickness of the foil layers in a direction perpendicular to their surface. Consequently, the current density, which is fairly uniform over the whole foil surface, can be expressed as the current through a single module divided by the large surface area of the foil layer which constitutes the cross-sectional area of the current path. It is, therefore, *low* even at high charge or dischage rates. In the case of many modules in parallel, the total battery current is, of course, as many times as large as the current through a single module without increase of the current density.

At a low current density there is not only a low heat development per unit area, but also a low level of all the forces of erosion. Furthermore, they cannot penetrate deeply as the foil layer has only a minute thickness. Perhaps the most important of the the latter is the low rate of development of gas bubbles per unit area which will be generated mainly on the foil surface and can escape freely. A low current density considerably reduces the rate of depletion of strength of the electrolyte near the foil surface.

The low current density, as a result of the large surface area, thus confers on the foil battery the benefits of safety and freedom from damage by the products and effects of chemical reactions often proceeding at high intensity in other systems.

Low internal resistance: The electrical resistance of a module is low because

• the length of path through the minute thickness of the foils and through the electrolyte is small
• the conductivity of the foils is high
• the foils are kept under endwise pressure over their whole area which is the very large cross-sectional area of the current path.

This is yet a further reason why the foil battery will allow high rates of charge and discharge without overheating and consequent damage to the battery.

Minimising damage by mechanical force: It might well be that foils have never been taken seriously as elements in a battery in view of their usual association with flimsiness. The invention overcomes the difficulty because it laminates all the foils and thin materials together and packs the laminates under elastic endwise pressure in a pile. This construction minimises the possibility of damage by vibration or the shedding of active material through chemical erosion or from other causes, and ensures the stability of the whole arrangement which is the battery.

Minimising progressive passivity: The other known major factor limiting the life of batteries is the progressive passivity of the active material. This is ascribed to the progressive envelopment of the active particles by insulating skins which are not reduced during the charging event due to loss of electric contact with the grid.

The change of particles losing contact is greatly reduced in the foil battery by virtue of the thinness of the active layers and by the fact that these layers are always held in pressure contact over their whole area with the barrier layers which take the place of the grids.

Limits of life: In addition to the above features, minimising the mechanical and chemical causes of a short life to batteries, the features referred to above must also be taken into account. Together they underline that the limits of life of the new battery are apparently going to be determined almost exclusively by their resistance to the attack of the electrolyte. An active continuation of the investigation of the materials to be used is therefore one of the main fields of work under the proposed development programme. At present, whereas it is not possible to give an accurate estimate of the expected life of the new battery, it can however be simply stated that it will be very long, substantially longer than that of the present type which it is designed to replace.

No compromise necessary: Previous attempts to use modules in battery construction have not been too successful. The foil-type battery can make full use of the advantages offered by modules because the proposed module is thin and flexible, the pile pressure elastic, there are no connections to be fixed and there is no container. As the module would be available at low enough price to render the maintenance by spare modules and the discarding of a damaged module free from financial objections, and as the versatility of the pile arrangement can cover many types of batteries, a considerable simplification of manufacturing, storing and handling becomes possible.

The module constructions also permit schemes for easy and efficient cooling.

Charging from any mains outlet: An important advantage of the small capacity modules is that they can be connected in parallel to give low voltage and large current for discharge while with a single switch they can be connected at any time in series to equal mains voltage and can, for charging overnight or when time is not pressing, take a low current for charging directly through a small rectifier without a transformer. Concerning rapid charging see below.

Although weight saving is not of prime important for the electric car, it is not to be overlooked that the weight of the new battery will be less than the weight of the present commercial type of equal capacity for the following reasons:

The active mass: Almost the whole active mass of the foil-like electrodes takes part in the chemical reactions during charge and discharge while in thick plates a very substantial surplus mass must be provided. No practical battery design for repeated cycles of charge and discharge has as yet been found to come as near to the minimum mass required by Faraday's law as the foil battery.

Grids and components: Almost all heavy chemically nonactive components have been eliminated. The foil battery has no container, no grids, no connectors, nor lugs, plugs, terminals or replacements of same. These also constitute a substantial part of the present weight of the conventional battery.

The new materials: The new materials (barrier foils, metal foils, insulating tapes, potting compounds) are neither bulky nor heavy. Thus, the weight of a foil battery will be much smaller than that of a conventional battery of equal capacity. Furthermore, for many applications, the safe high charge and discharge rate of the foil battery will permit the use of a foil battery of much smaller capacity than that of the conventional battery. This will, of course, mean a further saving to the user in weight and in space.

The impact which the above features are bound to make on cost is very substantial. It may be viewed under the following headings.

Lower cost of raw materials: This is based on the facts pointed out under 'Savings in weight and materials' above, which can be summarised as a substantial reduction of the amount of materials used which are the same as or as cheap as those used in the conventional battery.

Lower production cost by almost complete automation: One of the major objects of the design of the foil battery has been to render it possible to produce the whole module—not just plates or parts—completely automatically from long continuous rolls of material in a machine which is relatively simple, not requiring an excessive capital outlay. The consequent high output, coupled with a significant reduction in labour, will result in very low running cost.

Smaller batteries: In many instances, large capacity conventional batteries are used because large discharge currents are required over short times and not because the large capacity is of itself essential. The foil-type battery makes discharge rate to a great extent independent of capacity and much smaller, and therefore cheaper batteries will in many cases supply the need.

Long life: For reasons already given the foil battery is less prone to mechanical and chemical damage and deterioration. This is particularly the case when subjected to large discharge currents and mechanical shock. It should be noted that all the above reductions in cost are not made at the expense of the other important advantages of the foil-type battery, such as flexibility and low weight.

The only basically new material required for the lead-acid type of my bipolar foil battery was that of the barrier film. It posed a scientific problem which needed investigation in a top electrochemical laboratory starting with the study of various promising materials. Analysing the results, I concluded that the impermeable carbon I needed was characterised by its surface having the very minimum of porosity. This is obtained by firing it at very high temperature. Pyrolitic carbons and as I learned from the literature—tube material used in atomic reactors should therefore be suitable. That proved to be the case and the solution to the scientific problem.

Although the foil battery was not invented for the electric car, its basic capability, or being charged extremely quickly, could render a scheme for electric cars used essentially for local traffic, for runs in town and over smaller distancés, highly practical and promising.

Charging of a foil battery could be effected in a few minutes from high current mains outlets, to be made available every few miles at or near existing petrol stations, the battery having a capacity to drive a light type of two seater vehicle, say, 50 miles. For the larger car—still very much smaller than the new 'small' US models—a hybrid type could be used which comprises an equally small foil battery plus a 12 V petrol or steam driven turbogenerator on longer journeys and when no charging facilities are in easy reach. The generator would keep the capacity of the battery up.

Notes

1: The Beginnings in Vienna

1. For a general discussion of electrification in Germany see Thomas P. Hughes, *Networks of Power: Electrification in Western Society* (Baltimore and London: The Johns Hopkins University Press, 1983), especially pp. 315–19.

2. The problems of working at a distance from head office in a growing multinational company affected most industrial sectors; for another example, in the oil industry, see R. W. Ferrier, *The History of the British Petroleum Company. Volume 1: The Developing Years, 1901–1932* (Cambridge: Cambridge University Press, 1982), pp. 429–30.

3. The foreign exchange crisis was not, of course, restricted to Serbia. For a discussion of the problem across Europe, with particular reference to the impact of technical development see David S. Landes, *The Unbound Prometheus: Technological Change and Industrial Development Western Europe from 1750 to the Present* (Cambridge: Cambridge University Press, 1969, pp. 359–92.

4. Stereoscopic television had been discussed as early as 1928. See Albert Abramson, *Electronic Motion Pictures: A History of the Television Camera* (Berkeley, Calif.: University of California Press, 1955), pp. 40.

5. Marconi's interest in television is described in William J. Baker, *A History of the Marconi Company* (London: Methuen, 1970), pp. 256–66.

2: Early Experiences in England

1. There is a large literature on the early history of the telephone and the telegraph in the United States and Europe. For a wide-ranging analysis of the impact of telecommunications see Ithiel de Sola Pool, *The Social Impact of the Telephone* (Cambridge, Mass.: MIT Press, 1977), and the many works there cited. See also Robert Sobel, *ITT: The Management of Opportunity* (New York: Truman Talley Books, 1982), which tells the story of that firm's evolution and contributions to the development of telecommunications.

2. Views of the industry at the time may be read in *The Radio Industry: The Story of its Development*, introduced by David Sarnoff, (Chicago: A. W. Shaw Company, 1928: a series of lectures given at the Graduate School of Business Administration, Harvard). See also W. Rupert MacLaurin, *Invention and Innovation in the Radio Industry* (New York: Macmillan, 1949).

3. The approach of outsiders to large corporations has always been fraught with difficulties. As pointed out already companies are aware that agreeing to develop an invention brought in from outside can undermine their own R & D departments, or disrupt the equilibrium of the firm in any number of ways. Departments within large corporations usually work to budgets planned well in advance, and it can be very difficult for an outsider to break into this system. By adopting a rigid attitude the organisation runs the risk of missing an important discovery or development, but

from a cost accounting point of view that risk must be worth running. The finance department would undoutedly argue that the cost in terms of hours wasted talking to every inventor who came along could not be outweighed by the very rare occurrence whereby an important or money-spinning idea is forthcoming.

4. The situation of this company around this time is outlined in *British Thomson-Houston Progress in 1937* (Rugby: British Thomson-Houston, 1937).

5. This system is explained in Albert Abramson, *Electronic Motion Pictures: A History of the Television Camera* (Berkeley, Calif.: University of California Press, 1958), pp. 98–100.

6. *Ibid.* 19–20.

3: Printed Circuits: Origins and Early Applications

1. See Editors of *Electronics. An Age of Innovation: The World of Electronics 1930–2000* (New York: McGraw-Hill, 1981), chapter 4: "At War".

2. The growth of electronics since the Second World War is discussed in E. Braun and S. MacDonald, *Revolution in Miniature: The History and Impact of Semiconductor Electronics* (Cambridge: Cambridge University Press, 1978), from chapter 3 onwards. See also G. W. A. Dummer, *Electronic Inventions and Discoveries*, 3rd. Revised Edition (Oxford: Pergamon Press, 1983), especially chapter 3.

3. Whittle's story is told in Charles Harvard Gibbs-Smith, *Aviation: An Historical Survey From its Origins to the End of World War II* (London: HMSO, 1970), p. 196 and passim. See also Frank Whittle, *Jet: The Story of a Pioneer* (London: Frederick Muller, 1953).

4. A detailed account of thinking on the division of the labour process may be found in Harry Braverman, *Labour and Monopoly Capital: The Degradation of Work in the Twentieth Century* (New York and London: Monthly Review Press, 1974).

5. As described in Paul Eisler, *The Technology of Printed Circuits: The Foil Technique in Electronic Production* (London: Heywood & Company, 1959), chapter 3: "The Principles of Method Selection and the Foil Technique."

6. British Patent numbers 639 111, 639 178 and 639 179 respectively.

7. See Braun and MacDonald. *Op cit.* Chapter 11: "Reflections on an electronic age."

8. See M. D. Fagan, editor. *A History of Engineering and Science in the Bell System: National Service in War and Peace (1925–1975)* (Bell Telephone Laboratories, Limited, 1978), pp. 148–49.

4: Printed Circuits Take Off

1. The use of strain gauges is described in Paul Eisler, *The Technology of Printed Circuits: The Foil Technique in Electronic Production* (London: Heywood & Company, 1959), P. 273.

2. See E. Braun and S. MacDonald, *Revolution in Miniature: The History and Impact of Semiconductor Electronics* (Cambridge: Cambridge University Press, 1978), chapters 4, 5, and 9.

3. British patent No. 690 691. US patent No. 2747 977.

4. This work is described in Braun and MacDonald. *Op cit.*, chapter 4: "The Bell Laboratories".

5. For subsequent developments in the semiconductor industry in the United States and Britain see J. F. Tilton, *International Diffusion of Technology: The Case of Semi-conductors* (Washington DC: Brookings Institute, 1971). Also A. M. Golding, *The Semiconductor Industry in Britain and the United States: A Case Study in Innovation* (D. Phil. Thesis: University of Sussex, 1971). E. Sciberras, "The UK Semiconductor Industry" in *Technical Innovation and British Economic Performance,* edited by Keith Pavitt (London: Macmillan, 1980), pp. 282–96.

6. On the history of Pye see *The Story of Pye* (London: Pye Limited, c. 1960). There is also information on this company's move into the radio industry in M. J. G. Cattermole and A. F. Wolfe, *Horace Darwin's Shop: A History of The Cambridge Scientific Instrument Company 1878–1968* (Bristol and Boston: Adam Hilger, 1987), pp. 55–56.

7. The problem of raising capital to fund manufacturing capability or expansion of facilities was to recur regularly, as will be discussed in subsequent chapters. As a general problem for small firms the issue is discussed in Ray Oakey, *High Technology Small Firms: Regional Development in Britain and the United States* (London: Frances Pinter, 1984), chapter 9: "Finance and Innovation".

8. C. Brunetti and R. W. Curtis, *Printed Circuit Techniques* (Washington DC: National Bureau of Standards, Circular 468, 15 November 1947).

9. British Patent numbers 639 111, 639 178 and 639 179.

10. See C. Brunetti and R. W. Curtis, "Printed Circuit Techniques" *Proceedings of the Institution of Radio Engineers* 86, No. 1 (January 1948), pp. 121–61.

11. The impact of market conditions on innovatory activity is given a very thorough analysis in David C. Mowery and Nathan Rosenberg, "The Influence of Market Demand upon Innovation: A Critical Review of Some Recent Empirical Studies" *Research Policy* 8 (April 1979): 103–53. Reprinted in Nathan Rosenberg, *Inside the Black Box: Technology and Economics* (Cambridge: Cambridge University Press, 1982, pp. 193–241.

12. Eisler's experiences within a small organisation, and subsequently working on behalf of his own company, at the receiving end of the actions of larger and more powerful institutions are among the most revealing to emerge from his autobiography. For the business or industrial historian it is often extremely difficult to gain access to this kind of evidence. Ed.

13. British Patent No. 700 452: "Telephone Exchange Equipment."

14. The whole question of the political nature of technical artifacts is reviewed in Langdon Winner, "Do Artifacts Have Politics?" *Daedalus* 109 (1980): 121–36. Reprinted in *The Social Shaping of Technology,* edited by Donald MacKenzie and Judy Wajcman. (Milton Keynes and Philadelphia: Open University Press, 1985), pp. 26–38.

15. Taking on a powerful multinational with secure financial resources and expertise in the vital areas of marketing and the legal system could only have ended in disaster, as the corporation in question would be bound to do its utmost to protect itself. For instance, the history of the computer giant IBM is littered with examples of how it used its comparatively powerful situation to fight off lowlier competition in the market place. See Robert Sobel, *IBM: Colossus in Transition* (New York: Truman Talley Books, 1981; and London: Sidgwick & Jackson, 1984), especially Part II: "The Computer Wars".

5: Adventures Abroad

1. See the list of US patents in Paul Eisler, "Reflections of My Life as an Inventor" Part 3. *Circuit World* 11, No. 3 (1985), reprinted at the end of chapter 10.

2. On the British and United States patent systems see respectively Klaus Boehm and Aubrey Silberston, *The British Patent System* (Cambridge: Cambridge University Press, 1967). R. E. Brink, D. G. Cripple and H. Hughesdon, *An outline of US Patent Law.*

3. C. Brunetti and R. W. Curtis, *Printed Circuit Techniques* (Washington DC: National Bureau of Standards, Circular 468, 15 November 1947).

4. On the history of the relative industrial performance of Britain and the United States see David C. Mowery, "Firm Structure, Government Policy and the Organisation of Industrial Research: Great Britain and the United States, 1900–1950" *Business History Review* 58 (Winter 1984): 504–31. See also *Idem,* "Innovation, Market Structure, and Government Policy in the American Semiconductor Electronics Industry: A Survey." *Research Policy* 12 (1983), pp. 183–97.

6: Enter the NRDC

1. This organisation was created by the 1945–50 Labour Government, following the 1948 Development of Inventions Act. The background to its establishment, its role and aims are outlined in S. T. Keith, "Inventions, Patents and Commercial Development from Governmentally Financed Research in Great Britain: The Origins of the National Research Development Corporation" *Minerva* 19 (1981), pp. 92–122.

2. The NRDC has also been accused of mishandling other important technical developments, notably in the early British computer industry when the NRDC tried to negociate with two British firms, Elliott Brothers and Ferranti. This episode, which was unsatisfactory to all parties concerned, is described by John Hendry, "Prolonged Negotiations: The British Fast Computer Project and the Early History of the British Computer Industry" *Business History* 26, No. 3 (November 1984), pp. 280–306. See also P. Drath, *The Relationship Between Science and Technology: University Research and the Computer Industry, 1945–1962* (Ph.D. Thesis: University of Manchester, 1973).

3. When companies face financial problems one of the first areas to be cut is very often expenditure on R&D. The wisdom of such a move is debated at length and the arguments revolve around R&D as long-term investment or R&D for short-term results: whether R&D is a vital function of the firm or an expensive luxury to be funded when profits allow. For the small firm the dilemma is clearly set out in Ray Oakey, *High Technology Small Firms: Innovation and Regional Development in Britain and the United States* (London: Frances Pinter, 1984), pp. 92–96 and 125–27.

4. The movement of technical experts within industry can be an important source of industrial innovation and is a key element in technology transfer, as has been recognised for some time. In 1965 Don D. Price wrote that ". . . property may be coming a less influential form of capital than brains." *The Scientific Establishment* (Cambridge, Mass.: Harvard University Press, 1965),: 34. Quoted in Daniel Shimshoni, "The Mobile Scientist in the American Instrument Industry" *Minerva* 8 (1970), pp. 59–89. In this article Shimshoni explores in detail the impact of the movement of "brains" and confirms Price's observation.

5. See the list of patents.

6. Paul Eisler, "Printed Circuits: Some General Principles and Applications of the Foil Technique" *Journal of the British Institution of Radio Engineers* (November 1953), pp. 523–41.

7. There are many discussions of Edison as an inventor. See, for example, Frank L. Dyer and Thomas C. Martin, *Edison: His Life and Inventions.* 2 Volumes (New

York: Harper & Bros, 1930). Also Thomas P. Hughes, *Thomas Edison: Professional Inventor* (London: HMSO, 1977).

8. British Patent No. 690691: "Multilayer Materials for Printed Circuits."

9. See Paul Eisler, *The Technology of Printed Circuits: The Foil Technique in Electronic Production*. (London: Heywood & Company, 1959), chapter 17.

10. Inefficiency and inexperience were accusations often levelled at the NRDC by private inventors. See S. T. Keith, *op. cit.,* pp. 121–22.

11. As mentioned briefly in Paul Eisler, "Reflections on My Life as an Inventor" Part 1. *Circuit World* 11, No. 1 (1984). My aim for "Electrolimb" was to equip artificial limbs with electrical elements which are sensitive to touch, pressure and temperature, using the technology of printed circuitry.

12. While the general view of the NRDC, and particularly that held by the independent inventor, seems to be in accord with this, the Corporation did have a measure of success, in its support of biochemical research, as described in S. T. Keith, *op. cit.,* pp. 117–19.

13. Tension between the technically minded instigators of innovation and the people looking after the managerial—and especially financial—aspects of development is not restricted to the negociations of the NRDC during the 1950s. More recently, and perhaps more dramatically, events in Silicon Valley on the United States West Coast illustrate how the people with power are those with finance. Venture capitalists, so necessary to the success of that region will, if they think fit, bring in completely new management and move the inventors aside. See Judith K. Larsen and Everett M. Rogers, *Silicon Valley Fever: Growth of High Technology Culture* (London: George Allen & Unwin, 1984). Also, Oakey, *op. cit.,* pp. 140–41.

7: Freelancing

1. See the discussion on this issue in the previous chapter.

2. While the waste in large organisations is often apparent, the place of giant corporations in the modern economy of Western nations is both powerful and important, and is the result of a prolonged historical process. For the history of large corporations in Britain see S. Prais, *The Evolution of Giant Firms in Britain* (Cambridge: Cambridge University Press, 1976). Also, Leslie Hannah, *The Rise of the Corporate Economy,* second edition (London and New York: Metheun, 1983). The classic texts for the United States are Alfred D. Chandler Jr., *Strategy and Structure: Chapters in the History of American Industrial Enterprise* (Cambridge, Mass.: MIT Press, 1962). *Idem, The Visible Hand* (Cambridge, Mass.: Harvard University Press, 1977).

3. Paul Eisler, *The Technology of Printed Circuits: The Foil Technique in Electronic Production* (London: Heywood & Company. New York: Academic Press. 1959). A German translation was subsequently published by Carl Henser Verlag in Munich. The book was well received; a major science review journal described it as " . . . the most comprehensive treatment of the subject generally available, and it is clearly destined to become the standard work of reference on printed circuit technology." *Nature* No. 4674 (30 May 1959).

4. These techniques were being developed by mechanical and civil engineers as a means of analysing stresses in large structures, and I was fortunate to be able to discuss my ideas with engineers working on this type of research at University College, London.

5. Paul Eisler, "The Foil Battery." *Electric Vehicle Developments* (December 1980): 12–14. Reprinted in Appendix 4.

6. 3M is features in a more recent discussion of innovation, in which the main

problems analysed are those faced by a firm which adopts a new product and needs to convince potential buyers of its use. In P. Ranganath Nayak and John Ketteringham, *Breakthroughs.* (Rawson Associates, 1986), 3M's development and marketing of "Post-it" pads are used as one example of the difficulties an organisation still faces after doing what economic advisers keep urging, and innovating.

7. See Appendix 1.

8. The whole impact of domestic appliances on the life of the housewife is analysed in Ruth Schwartz Cowan, "The Industrial Revolution in the Home" and "How the Refrigerator got its Hum" in *The Social Shaping of Technology*, edited by Donald MacKenzie and Judy Wajcman. (Milton Keynes and Philadelphia: Open University Press, 1985), pp. 181–201 and 202–18.

9. Again it is clear here that the apparatus for financing small scale science-based industry simply did not exist in Britain in the late 1950s. Nor has it been created to any great extent since. See Ray Oakey, *High Technology Small Firms: Innovation and Regional Development in Britain and the United States* (London: Frances Pinter, 1984), pp. 125–31. See also the Wilson Committee, *The Financing of Small Firms: Interim Report of the Committee to Review the Functioning of the Financial Institutions* (London: HMSO, Cmnd 7503, 1979).

10. See list of Pual Eisler's patents at the end of chapters 10.

11. The role of the business entrepreneur within the process of establishing small firms and the promotion of innovation is another much considered issue; general agreement exists that the British have lacked and still lack entrepreneurial drive when compared with other Western nations. See David S. Landes, *The Unbound Prometheus: Technological Change and Industrial Development in Western Europe from 1750 to the Present* (Cambridge: Cambridge University Press, 1969), p. 337; also W. B. Walker, "Britain's Industrial Performance 1850–1950" in *Technical Innovation and British Economic Performance*, edited by Keith Pavitt (London: Macmillan, 1980), pp. 19–37. By comparison in the United States a number of inventor-entrepreneurs have had considerable success. See Thomas P. Hughes, *Elmer Sperry: Inventor and Engineer* (Baltimore and London: The Johns Hopkins University Press, 1971). *Idem*, "Inventors: The Problems They Choose, the Ideas They Have and the Inventions They Make" in *Technical Innovation: A Critical Review of Current Knowledge*, edited by P. Kelly and M. Kranzberg, (San Francisco: San Francisco Press, 1978), pp. 166–82.

12. Given all these reasons which a large organisation might offer for refusing to take on new ideas from outside it is perhaps not surprising that so many are turned down. For example, the NRDC supported very few of the more than 19,000 inventions submitted by private individuals, as explained by S. T. Keith, "Inventions, Patents and Commercial Development from Governmentally Financed Research in Great Britain: The Origins of the National Research Development Corporation" *Minerva* 19 (1981): 92–122. 121–22. No doubt any number of potentially successful ideas will have been missed in this way. But perhaps one of the greatest missed opportunities occurred in California when the computer company Hewlett-Packard turned down the idea for a microcomputer offered to them by Steve Wozniak and Steve Jobs; the two men went on to create Apple Computers, one of the greatest success stories of Silicon Valley. See Judith K. Larsen and Everett M. Rogers, *Silicon Valley Fever: Growth of High Technology Culture* (London: George Allen & Unwin, 1984), chapter 1.

8: The Story of Space Heating

1. British Patent No. 639 178: "Foil Technique of Printed Circuits."

2. See above. Chapter 6, note 11.

3. The ability of firms to diversify successfully also plays an important part in their survival. Such ability is one of the characteristics of high technology industry: for example recently a number of major oil companies have invested in other energy-related areas as a way of ensuring the future as oil as a resource dwindles. It is, however, far from clear that other, more traditional sectors of industry are capable of successful diversification, as this episode in Paul Eisler's career confirms. Ed.

4. This illustrates yet another problem in-built into the establishment of new products: the time-lag which is nearly always involved. The time needed to get from an idea to a working product or process is an aspect of innovation which some economists have tried to incorporate into their models. See, for example, C. F. Carter and B. R. Williams, *Industry and Technical Progress: Factors Governing the Speed of Application of Science* (London: Oxford University Press, 1957). But it is extremely difficult to model this particular problem and, as Paul Eisler experienced, if the time available does not tally with the time needed failure may ensue. Ed.

5. It is possible to speculate that one of the reasons why Eisler's initiatives often met with untimely ends is that the type of work in which he was involved— inventions based on novel applications of electric currents in unusual circuits—has all the appearance of the high technology side of industry; whereas the companies with which he was trying to negociate were in traditional areas. As a result he was faced with coping with a difference in industrial culture which was usually too great to overcome. Ed.

6. Sometimes a decision not to innovate can be rooted in reasons quite as legitimate as a decision to break new ground. See *Industrial Policy and Innovation*, edited by Charles Carter (London: Heinemann, 1981), chapter 3: "Reasons for not Innovating". See also Donald Schon, "The Fear of Innovation" in *Science in Context: Readings in the Sociology of Science* (Milton Keynes: Open University Press, 1982), pp. 290–302.

7 British Patent number 639 658: "Tin Printing."

8. Once again is attention drawn to the problem of raising funds for new initiatives. See Chapter 8, note 4, and the works there cited.

9. The story of the discovery of North Sea gas is told in Trevor I. Williams, *A History of the British Gas Industry* (Oxford: Oxford University Press, 1981), chapter 16.

10. See *Ibid*, chapter 17. Also Charles Elliott, *The History of Natural Gas Conversion in Great Britain* (Royston: Cambridge Information and Research Services Ltd., 1980).

11. Williams, *op. cit.*, chapter 20.

12. The story which unfolded resembled a nightmare more and more as time passed. There was persistent opposition from one member of the committee whose job it was to vet proposals. The time delay brought financial difficulties, a change of management and, finally, our enforced withdrawal.

9: Expansion and Final Success

1. As discussed previously (Chapter 4, note 15) the prospect of competing with large organisations can be daunting; and the chances of success are slim.

2. On electricity tariffs see Leslie Hannah, *Engineers, Managers and Politicians* (London: Macmillan, 1982).

3. All firms encountering initial success will reach a point at which a decision must be made about future growth. In many cases—for example in the fast growth high technology industry of Silicon Valley—one of the main points is to expand to

the situation from which flotation on the stock market is possible, at vast profit to the initial investors. But this will always mean relinquishing control and, in Britain more particularly, it has been more usual for those in control of a firm to opt to retain it.

4. British Patent number 1 333 049: "Improvements in and relating to Heating Panels"; US Patent number 3 736 404: "Combined Demisting and and Defrosting Heating Panel for Windows and Other transparent Areas."

5. See Appendix 2 for further details.

6. See A. E. Canham, *Electricity in Horticulture* (London: Macdonald, 1964).

"When one considers the amount of time, money and energy that have to be spent by growers and research workers in the search for the ideal environment in which to grow plants, it is little short of incredible that only a very small proportion of this effort has been devoted to the consideration of the subterranean microclimate. The nutrient status of the soil has been given due recognition: the importance of an adequate supply of water is well known—although the actual needs of the plant for optimal growth are still somewhat enigmatic. The question of temperature of the soil, however, has received scant attention and with only a few notable exceptions little has been done to find out how important this factor is in the life of plants.

"For optimal growth, all growth factors must be optimal; the most important factor at any one time is the one which is sub-optimal. This acts as a brake on the rate of growth and in some circumstances soil temperature can be that brake. If it is, electrical soil warming is a convenient and effective means of removing the brake and ensuring unimpeded progress in growth."

7. See Appendix 3.

8. See Appendix 4.

9. Of the many large organisations dominating the contemporary business and industrial worlds ICI has one of the better reputations, for success and management. The path the company has followed to reach such a stage is recounted in W. J. Reader, *Imperial Chemical Industries: A History* Volume 2, *The First Quarter-Century* (Oxford: Oxford University Press, 1975).

10. See Appendix 3.

11. As described in Appendix 4.

12. Menlo Park is, of course, now one of the main towns of Silicon Valley, and while it may be too easy to ascribe success there simply to its geographical location, it is possible that this was one of the factors involved.

Bibliography

Abramson, Albert. *Electronic Motion Pictures: A History of the Television Camera.* Berkeley, Calif.: University of California Press, 1955.

Aitken, Hugh G. J. *Syntony and Spark: The Origins of Radio.* New York: Wiley, 1976.

Baker. W. J. *A History of the Marconi Company.* London: Methuen, 1970.

Boehm, Klaus, and Silberston, Aubrey. *The British Patent System.* Cambridge: Cambridge University Press, 1967.

Braun, E., and Macdonald, S. *The History and Impact of Semiconductor Electronics.* Cambridge: Cambridge University Press, 1978.

Braverman, Harry. *Labour and Monopoly Capital: The Degradation of Work in the Twentieth Century.* New York and London: Monthly Review Press, 1974.

Brink, R. E., Cripple, D. G., and Hughesdon, H. *An Outline of US Patent Law.* New York: Interscience Publishers, 1959.

British Thomson-Houston. *British Thomson-Houston Progress in 1937.* Rugby: British Thomson-Houston, 1937.

Brock, G. W. *The Telecommunications Industry.* Cambridge, Mass.: Harvard University Press, 1981.

Brunetti, Cledo, and Curtis, Roger W. *Printed Circuit Techniques.* Washington DC: National Bureau of Standards, Circular 468, 15 November 1947.

———. "Printed Circuit Techniques." *Proceedings of the Institution of Radio Engineers* 86, No. 1 (January 1948): 121–61.

Buxton, N. K. "The Role of the 'New Industries' in Britain During the 1930s: A Reinterpretation." *Business History Review* 49 (1975): 312–36.

Canham, A. E. *Electricity in Horticulture.* London: Macdonald, 1964.

Carter, Charles. Editor. *Industrial Policy and Innovation.* London: Heinemann, 1981.

Carter, C. F., and Williams, B. R. *Industry and Technical Progress: Factors Governing the Speed of Application of Science.* London: Oxford University Press, 1957.

Cattermole, M. J. G., and Wolfe, A. F. *Horace Darwin's Shop: A History of the Cambridge Scientific Instrument Company* 1878–1968. Bristol and Boston: Adam Hilger, 1987.

Chandler, Alfred D. Jr. *Strategy and Structure: Chapters in the History of the American Industrial Enterprise.* Cambridge, Mass.: MIT Press, 1962.

———. *The Visible Hand.* Cambridge, Mass.: Harvard University Press, 1977.

Constant, Edward W. II. *The Origins of the Turbojet Revolution.* Baltimore and London: The Johns Hopkins University Press, 1980.

Cowan, Ruth Schwartz. "How the Refrigerator got its Hum." In *The Social Shaping of Technology.* Edited by Donald MacKenzie and Judy Wajcman. Milton Keynes and Philadelphia: Open University Press, 1985.

———. "The Industrial Revolution in the Home." In *The Social Shaping of Technology.* Edited by Donald MacKenzie and Judy Wajcman. Milton Keynes and Philadelphia: Open University Press, 1985.

Dineen, G. P., and Frick, F. C., "Electronics and National Defense: A Case Study." *Science* 195 (March 1977): 1151–55.

Drath, P. *The Relationship Between Science and Technology: University research and the Computer Industry, 1945–1962.* Ph.D. Thesis: University of Manchester, 1973.

Dummer, G. W. A. *Electronic Inventions and Discoveries.* 3rd Revised Edition. Oxford: Oxford University Press, 1983.

Dyer, Frank L., and Martin, Thomas C. *Edison: His Life and Inventions.* 2 Volumes. New York: Harper & Bros, 1930.

Editors of *Electronics. An Age of Innovation: The World of Electronics 1930–2000.* New York: McGraw-Hill, 1981.

Eisler, Paul. "Printed Circuits: Some General Principles and Applications of the Foil Technique." *Journal of the British Institution of Radio Engineers* 13, No. 11 (November 1953): 523–41.

———. "Reflections of My Life as an Inventor." *Circuit World* 11, Nos. 1–3 (1984–85).

———. *The Technology of Printed Circuits: The Foil Technique in Electronic Production.* London: Heywood & Company, 1959. New York: Academic Press, 1959.

Elliott, Charles. *The History of Natural Gas Conversion in Great Britain.* Royston: Cambridge Information and Research Services ltd., 1980.

Fagan, M. D. *A History of Engineering and Science in the Bell System: National Service in War and Peace (1925–1975).* Bell Telephone Laboratories Ltd., 1978.

Ferrier, R. W. *The History of the British Petroleum Company.* Volume 1. *The Developing Years, 1901–1932.* Cambridge: Cambridge University Press, 1982.

Forester, Tom. Editor. *Microelectronics Revolution: The Complete Guide to the New Technology and its Impact on Society.* Oxford: Blackwell, 1980.

Gibbs-Smith, Charles Harvard. *Aviation: An Historical Survey From its Origins to the End of World War II.* London: HMSO, 1970.

Golding, A. M. *The Semiconductor Industry in Britain and the United States: A Case Study in Innovation.* D. Phil. Thesis: University of Sussex, 1971.

Graduate School of Business Administration, Harvard. *The Radio Industry: The Story of its Development.* Lectures given to the Business School, introduced by David Sarnoff. Chicago: A. W. Shaw Company, 1928.

Hannah, Leslie. *Engineers, Managers, and Politicians.* London: Macmillan, 1982.

———. *The Rise of the Corporate Economy.* 2nd Edition. London and New York: Methuen, 1983.

Hendry, John. "Prolonged Negotiations: The British Fast Computer Project and the Early History of the British Computer Industry." *Business History* 26, No. 3 (November 1984): 280–306.

Hughes, Thomas P. *Elmer Sperry: Inventor and Engineer.* Baltimore and London: The Johns Hopkins University Press, 1971.

———. "Inventors: The Problems they Choose, the Ideas they Have and the Inventions they Make." In *Technological Innovation: A Critical Review of Current Knowledge.* Edited by P. Kelly and M. Kranzberg. San Francisco: San Francisco Press, 1978.

———. *Networks of Power: Electrificaton in Western Society.* Baltimore and London: The Johns Hopkins University Press, 1983.

———. *Thomas Edison, Professional Inventor.* London: HMSO, 1976.

Jack, Mervyn A. Editor. *The Impact of Microelectronics Technology.* Edinburgh: Edinburgh University Press, 1982.

Jones, R., and Marriot, O. *Anatomy of a Merger: A History of GEC, AEI, and English Electric.* London: Jonathan Cape, 1970.

Jones, Trevor W. Editor. *Microelectronics and Society.* Milton Keynes: Open University Press, 1980.

Keith, S. T. "Inventions, Patents and Commercial Development from Governmentally Financed Research in Great Britain: The Origins of the National Research Development Corporation." *Minerva* 19 (1981): 92–122.

Landes, David S. *The Unbound Prometheus: Technological Change and Industrial Development in Western Europe from 1750 to the Present.* Cambridge: Cambridge University Press, 1969.

Langrish, J., Gibbons, M., Evans, W. G., and Jevons, F. R. *Wealth From Knowledge: A Study of Innovation in Industry.* London: Macmillan, 1972.

Larsen, Judith K., and Rogers, Everett M. *Silicon Valley Fever: Growth of High Technology Culture.* London: George Allen & Unwin, 1984.

Lavington, Simon. *Early British Computers.* Manchester: Manchester University Press, 1980.

MacLaurin, W. Rupert. *Invention and Innovation in the Radio Industry.* New York: Macmillan, 1949.

Metropolis, N., Howlett, J., and Rota, Gian-Carlo. Editors. *A History of Computing in the Twentieth Century.* New York: Academic Press, 1980.

Mowery, David C. "Firm Structure, Government Policy and the Organization of Industrial Research: Great Britain and the United States, 1900–1950." *Business History Review* 58 (Winter 1984): 504–31.

———. "Industrial Research, 1900–1950." In *The Decline of the British Economy.* Edited by B. Elbaum and W. Lasonick. Oxford: Oxford University Press, 1986.

———. "Innovation, Market Structure, and Government Policy in the American Semiconductor Electronics Industry: A Survey." *Research Policy* 12 (1983): 183–97.

Mowery, David C. and Rosenberg, Nathan. "The Influence of Market Demand upon Innovation: A Critical Review of Some Recent Empirical Studies." *Research Policy* 8 (April 1979): 103–53. Reprinted in Nathan Rosenberg. *Inside the Black Box: Technology and Economics.* Cambridge: Cambridge University Press, 1982.

Nayak, P. Ranganath, and Ketteringham, John. *Breakthroughs.* Rawson Associates, 1987.

Oakey, Ray. *High Technology Small Firms: Innovation and Regional Development in Britain and the United States.* London: Frances Pinter, 1984.

Ogburn, William F., and Thomas, Dorothy. "Are Inventions Inevitable?" *Political Science Quarterly* 43 (1922): 83–98.

Pool, Ithiel de Sola. *The Social Impact of the Telephone.* Cambridge, Mass.: MIT Press, 1977.

Prais, S. *The Evolution of Giant Firms in Britain.* Cambridge: Cambridge University Press, 1976.

Price, Don K. *The Scientific Establishment.* Cambridge, Mass.: Harvard University Press, 1965.

Pye Limited. *The Story of Pye.* London: Pye Ltd., c. 1960.

Reader, W. J. *Imperial Chemical Industries: A History.* Volume 2. *The First Quarter-Century.* Oxford: Oxford University Press, 1975.

Reich, L. S. "Research, Patents, and the Struggle to Control Radio: A Study of Big Business and the Uses of Industrial Research." *Business History Review* 51 (1977): 208–55.

Rosenberg, Nathan. *Inside the Black Box: Technology and Economics.* Cambridge: Cambridge University Press, 1982.

Rothwell, R., and Zegveld, W. *Innovation and Small And Medium Sized Firms.* London: Frances Pinter, 1982.

Saul, S. B. "The Engineering Industry." In *The Development of British Industry and Foreign Competition.* Edited by D. H. Aldcroft. London: George Allen & Unwin, 1968.

Schallenberg, Richard H. "The Alkaline Storage Battery: A Case History of the Edison Method." *Synthesis* 1 (1972); 1–13.

Scherer, F. M. "Firm Size, Market Structure, Opportunity, and the Output of Patented Inventions." *American Economic Review* 55 (1965): 1097–1125.

Schmookler, Jacob. *Inventions and Economic Growth.* Cambridge, Mass.: Harvard University Press, 1966.

Schon, Donald. "The Fear of Innovation." *In Science in Context: Readings in the Sociology of Science.* Milton Keynes: Open University Press, 1982.

Sciberras, E. "The UK Semiconductor Industry." In *Technical Innovation and British Economic Performance.* Edited by Keith Pavitt. London: Macmillan, 1980.

Shimshoni, Daniel. "The Mobile Scientist in the American Instrument Industry." *Minerva* 8 (1970): 59–89.

Sobel, Robert. *IBM: Colossus in Transitition.* New York: Truman Talley Books, 1981. London: Sidgwick & Jackson, 1984.

———. *ITT: The Management of Opportunity.* New York: Truman Talley Books, 1982.

Soete, Luke. "Firm Size and Inventive Activity: The Evidence Reconsidered." *European Economic Review* 12 (1979): 319–40.

Technology Policy Unit. *The Impact of Microelectronics: A Review of the Literature.* London: Frances Pinter, 1981.

Tilton, J. F. *International Diffusion of Technology: The Case of Semi-conductors.* Washington DC: Brookings Institute, 1971.

Usher, Abbott Payson. *A History of Mechanical Inventions.* Revised Edition. Cambridge, Mass.: Harvard University Press, 1954.

von Hippel, Eric. "The Dominant Role of the User in Semiconductor and Electronic Subassembly Process Innovation." *IEEE Transactions on Engineering Management* EM-24, No. 2 (May 1977): 60–71

Walker, W. B. "Britain's Industrial Performance 1850–1950." In *Technical Innovation and British Economic Performance.* Edited by Keith Pavitt. London: Macmillan, 1980.

Whittle, Frank. *Jet: The Story of a Pioneer.* London: Frederick Muller, 1953.

Williams, Trevor I. *A History of the British Gas Industry.* Oxford: Oxford University Press, 1981.

Wilson Committee. *The Financing of Small Firms: Interim Report of the Committee to Review the Functioning of the Financial Institutions.* London: HMSO, Cmnd 7503, 1979.

Winner, Langdon. "Do Artifacts Have Politics?" In *The Social Shaping of Technology.* Edited by Donald MacKenzie and Judy Wajcman. Milton Keynes and Philadelphia: Open University Press, 1985.

Index